CHINA AND ASEAN

ISEAS–Yusof Ishak Institute (formerly the Institute of Southeast Asian Studies) was established as an autonomous organization in 1968. It is a regional centre dedicated to the study of socio-political, security and economic trends and developments in Southeast Asia and its wider geostrategic and economic environment. The Institute's research programmes are the Regional Economic Studies (RES, including ASEAN and APEC), Regional Strategic and Political Studies (RSPS), and Regional Social and Cultural Studies (RSCS).

ISEAS Publishing, an established academic press, has issued more than 2,000 books and journals. It is the largest scholarly publisher of research about Southeast Asia from within the region. ISEAS Publishing works with many other academic and trade publishers and distributors to disseminate important research and analyses from and about Southeast Asia to the rest of the world.

CHINA AND ASEAN

⋄⋄⋄ Energy Security, Cooperation and Competition ⋄⋄⋄

ZHAO HONG

First published in Singapore in 2015 by
ISEAS Publishing
30 Heng Mui Keng Terrace
Singapore 119614

E-mail: publish@iseas.edu.sg • Website: bookshop.iseas.edu.sg

All rights reserved. No part of this publication may be reproduced, stored in a retrieval system, or transmitted in any form or by any means, electronic, mechanical, photocopying, recording or otherwise, without the prior permission of the ISEAS–Yusof Ishak Institute.

© 2015 ISEAS–Yusof Ishak Institute, Singapore

The responsibility for facts and opinions in this publication rests exclusively with the author and his interpretation do not necessarily reflect the views or the policy of the publisher or its supporters.

ISEAS Library Cataloguing-in-Publication Data

Zhao, Hong.
China and ASEAN : Energy Security, Cooperation and Competition.
1. Energy policy—China.
2. Energy policy—Southeast Asia.
3. Energy consumption—China.
4. Energy consumption—Southeast Asia.
5. China—Foreign relations—Southeast Asia.
6. Southeast Asia—Foreign relations—China.
7. China—Foreign economic relations—Southeast Asia.
8. Southeast Asia—Foreign economic relations—China.
9. South China Sea—International status.
I. Title.
HD9502 C52Z631 2015

ISBN 978-981-4695-25-1 (soft cover)
ISBN 978-981-4695-26-8 (E-book PDF)

Typeset by International Typesetters Pte Ltd
Printed in Singapore by Markono Print Media Pte Ltd

CONTENTS

LIST OF TABLES AND FIGURES

Tables

Figures

LIST OF ABBREVIATIONS

ACE	ASEAN Center for Energy
ACIA	ASEAN Comprehensive Investment Agreement
AEMI	ASEAN Energy Market Integration
AIA	ASEAN Investment Area
APAEC	ASEAN Plans of Action for Energy Cooperation
ARF	ASEAN Regional Forum
ASCOPE	ASEAN Council on Petroleum
bcm	cubic metres
BCF	billion cubic feet
b/d	barrel per day
boe	barrels of oil equivalent
CNOOC	China National Offshore Oil Corporation
CNPC	China National Petroleum Corporation
COC	Code of Conduct
DOC	Declaration on the Conduct
EEZ	Exclusive Economic Zone
GW	gigawatt
IEEJ	Institute of Energy Economics
IPCC	Intergovernmental Panel on Climate Change

Ktoe	thousand tons of oil equivalent
Kb/d	thousand barrels per day
LNG	Liquefied Natural Gas
MMBtu	million British thermal units
Mtce	million tons of coal equivalent
Mtoe	million tons of oil equivalent
Mt	million tones
NDRC	National Development and Reform Commission
NOCs	National Oil Companies
ODA	Official Development Aid
OPEC	Organization of the Petroleum Exporting Countries
PLAN	People's Liberation Army Navy
PSC	Production Sharing Contracts
RCEP	Regional Comprehensive Economic Partnership
Sinopec	China National Petrochemical Corporation
TAGP	Trans-ASEAN Gas Pipeline
Tcf	trillion cubic feet
TPP	Trans-Pacific Partnership
Twh	terawatt-hour
UNCLOS	United Nations Convention on the Law of the Sea
ZoPFFC	Zone of Peace, Freedom, Friendship and Cooperation

ACKNOWLEDGEMENTS

In writing this book, I have fortunately had enormous support and assistance from various individuals and institutes. That I could eventually finish this humble book was due to many people who helped and supported me in my academic career over the past years. Foremost, I would like to thank ISEAS–Yusof Ishak Institute Director Ambassador Tan Chin Tiong and Deputy Director Ooi Kee Beng. Without their encouragement and support, it would not have been possible to undertake this research project in China and Southeast Asia. In undertaking my field research, I am specially indebted to the following institutes. In Yunnan, Yunnan Normal University arranged a trip for me to the border areas. For this, I am grateful to Professor Luo Huasong. In Xiamen University, I am indebted to the staff of the Research School of Southeast Asian Studies for their time and interviews. They are Professor Li Jinming, Professor Fan Hongwei, Professor Wang Qin, and Professor Wu Chongbo. In Singapore, EAI (East Asian Institute) gave me a special opportunity to conduct my research at its library. For this I am so grateful to Professor Zheng Yongnian and Dr Lam Peng Er. Heartfelt thanks also go to my colleagues and friends at ISEAS–Yusof Ishak Institute for their encouragement and collegiality: Mr Daljit Singh, Dr Terence Chong, Dr Malcolm Cook, Dr Ian Storey, Dr Francis E. Hutchinson, Dr Maxensius Tri Sambodo, Dr Lee Poh Onn, Dr Cassey Lee, and Dr Tang Siew Mun. Appreciation too goes to the Institute's editorial team, Mr Ng Kok Kiong and Ms Sheryl Sin Bing Peng for bringing this book into fruition in a timely manner. Many thanks also go to *Trends in Southeast Asia* for its permission to revise and publish the following articles: "The South China Sea and China–ASEAN relations" (parts of Chapter 6); and "China's quest for energy in Southeast Asia" (parts of Chapter 3). Parts of Chapter 5 (Case Study of Indonesia) appeared in a draft paper presented at an international workshop organized by ISEAS in May 2015.

Finally, without thanking my wife and daughter, the book cannot be complete. I am deeply indebted to my wife's endless support, encouragement, and understanding. Equally important is my daughter who has been always with me through *WeChat* and e-mail, sharing hardship and happiness. It is to you that I dedicate this book.

1

INTRODUCTION China, ASEAN, and the New Global Energy Order

CHINA'S RISE AND SOUTHEAST ASIA

This book has been motivated largely by the rapid rise of China and its consequent influence in the world, particularly in Southeast Asia. As an emerging power in East Asia, China is the main driver behind the geopolitical and economic reconfiguration that is taking place in Asia. Southeast Asia is one of the areas that is affected most directly by the rise of China. On the other hand, interactions between China and ASEAN (Association of Southeast Asian Nations) will, to a great extent, affect the future and prospect of the entire Asia, leading East Asia to become the world's new economic centre of gravity, and affect the world economic and energy map.

In East Asia, few relations have evolved as much as that between China and ASEAN.[1] The threat of China looms large in the history of the relations between China and Southeast Asian countries. In particular, Chinese support for local communist groups during the 1960s and its views of ASEAN as an anti-Chinese, anti-communist alliance created distrust and frictions in China–ASEAN relations.[2] China's siding with ASEAN during Vietnam's occupation of Cambodia contributed to re-establishing relations with Thailand, Malaysia, and the Philippines, but those between China and Brunei, Indonesia, and Singapore were not restored until 1990–91.[3] Mutual suspicion lingered through much of the 1990s, due to memories of its support for communist insurgencies and the Chinese tendency to dismiss these smaller countries as puppets of U.S. imperialism.

The end of the Cold War was a pivotal turning point in China–ASEAN relations. ASEAN and China began to perceive complementary advantages in closer cooperation. Southeast Asia had energy resources which China did not have; China was modernizing rapidly and could contribute to modernization in Southeast Asia. China embarked on a new policy that was more geared toward the East, with Southeast Asia as a major focus. China's "good neighbour policy" was aimed at strengthening regional relations so as to surround itself with benevolent states, which would allow China to focus on its economic development.[4] At the same time, post-Cold War uncertainties about the U.S. policies created new pressure on ASEAN to find other ways to stabilize and expand their relations with China. For the ASEAN states, the economic and political-security uncertainties surrounding the U.S. role in Southeast Asia made it especially important for ASEAN to engage China and improve relations as a kind of hedge against the possibility of further U.S. retrenchment.[5] The turning point for ASEAN's perceptual change about China from "China as a threat" to "China as an opportunity", many Chinese analysts believe, is the Asian financial crisis in 1997–98.[6] After that, the mutual interests and avenues of cooperation increased. The ASEAN states, which had developed certain values and norms to facilitate cooperation in a diverse religion-ethnic mix, began to conceive the possibility that China could be socialized into the "ASEAN Way". China was invited to participate as a dialogue partner in the ASEAN Regional Forum (ARF) and helped set up the ASEAN plus one and ASEAN plus three discussion forums.

In 2003, Zheng Bijian, Chair of the China Reform Forum, made a speech saying that the rise of a new great power oftentimes had led to great turbulence in the international system. One important reason was that the new great power usually tried to resort to wars to destroy the existing international system. Zheng explained that China shall adopt a different approach, which he later elaborated as: "transcending 'the traditional ways for great powers to emerge, as well as the Cold War mentality that defined international relations along ideological lines'."[7] China's strategy, according to Zheng, should be a peaceful rise, i.e. working for a peaceful international environment for China's development, and in turn safeguarding world peace with China's development. It turns out that Zheng's ideas are actually China's strategy. Both President Hu Jintao and Premier Wen Jiabao espoused the road of peaceful rise. In 2005 and 2011, respectively, the Chinese government published two white papers on peaceful development. President Xi Jinping also called for building a "community of shared interests" and a "community of shared destiny" which will provide the vision for realizing Asia's economic potential and achieving more durable security for Asia.[8] China aims to bind its interests more closely with the countries at its doorstep.

China is using trade and investment, confidence-building measures, and development assistance to establish itself as an important regional leader. This was reflected in China's proposed establishment of the China–ASEAN FTA (CAFTA), which came into force on 1 January 2010, and the Asian Infrastructure Investment Bank (AIIB), which is to be fully established by the end of 2015. CAFTA serves important political goals, especially in terms of confidence-building, as well as solidifying and further increasing its influence in the region. CAFTA grants China access to the energy and raw materials of resource-endowed Southeast Asia, as well as providing an increased market for Chinese products and capital within Southeast Asia. For ASEAN countries, CAFTA offers access to China's market and an opportunity to cash in on China's increased wealth and consumer spending.

For both parties, CAFTA serves to diversify China's and Southeast Asia's trade assay from the West. For example, the share of ASEAN's trade with the United States in its total trade decreased from 17.7 per cent in 2002 to 8.1 per cent in 2012, the share of its trade with EU-27 decreased from 13.6 per cent to 9.8 per cent; while the share of its trade with China in its total trade increased from 6 per cent to 13 per cent during the same period (see Table 1.1). CAFTA opens up new avenues and is expected to boost China–ASEAN trade alongside expanding intra-industry trade and increased investment flow between the two sides. China's trade with ASEAN as a whole grew about ninefold in just ten years from US$32 billion in 2001 to US$287.6 billion in 2011, and further to US$443.6 billion in 2013, with most ASEAN countries recording trade surpluses with China.[9] Chinese investment in Southeast Asia increased even more, from a meagre US$144 million in 2001 to US$6.1 billion in 2012,[10] and that includes only officially reported investments.

However, although China has been actively promoting the idea that it is engaged in a peaceful rise, the message has not been wholeheartedly embraced by its neighbours. Beginning in 2010, the mutual economic and social integration and socialization process began to encounter problems that could not be readily resolved. ASEAN countries find the economic importance of China has increasingly grown, but they still remain uncertain about Beijing's long-term intentions and the consequences of China's economic activities in Southeast Asia. They are particularly concerned that as China rises economically and militarily, its request for natural resources in some Southeast Asian resource-rich countries and the South China Sea

TABLE 1.1
Top Five ASEAN Trade Partners
(US$ million; %)

	2002	2003	2004	2005	2006	2007	2008	2009	2010	2011	2012
China	42,759 (6)	59,637 (7.6)	89,066 (8.3)	113,393 (9.3)	139.961 (10)	171,117 (10.6)	196,883 (10.4)	178,185 (11.6)	236,219 (11.8)	287,676 (12.1)	319,485 (12.9)
EU-27	97,056 (13.6)	101,364 (12.3)	131,543 (12.3)	140,237 (11.5)	160,332 (11.4)	186,719 (11.6)	208,291 (11)	171,785 (11.2)	214,091 (10.7)	240,248 (10.1)	242,599 (9.8)
Japan	97,587 (13.7)	113,400 (13.8)	143,263 (13.4)	153,834 (12.6)	161,780 (11.5)	173,062 (10.7)	214,400 (11.3)	160,863 (10.5)	218,963 (11)	255,048 (10.7)	262,884 (10.6)
South Korea	30,533 (4.3)	33,548 (4.1)	40,543 (3.8)	47,971 (3.9)	55,942 (4)	61,184 (3.8)	78,250 (4.1)	74,740 (4.9)	102,871 (5.1)	124,381 (5.2)	131,030 (5.3)
U.S.	104,954 (17.7)	117,885 (14.3)	135,864 (12.7)	153,918 (12.6)	161,195 (11.5)	179,068 (11.1)	186,242 (9.8)	149,572 (9.7)	192,295 (9.6)	196,502 (8.1)	200,027 (8.1)
Total	713,816 (100)	824,538 (100)	1,071,604 (100)	1,224,889 (100)	1,404,805 (100)	1,610,787 (100)	1,897,127 (100)	1,536,843 (100)	1,998,155 (100)	2,386,584 (100)	247,427 (100)

Source: ASEAN Trade Statistics Database.

might spark conflicts there. In the process of China's rise, the shortage of resources poses a big problem. The scarcity of natural resources available to support China's huge population and high economic growth rate is a big challenge that China must confront. The fact that China's oil, natural gas, copper, and aluminum resources in per capita terms amount to 8.3 per cent, 4.1 per cent, 25.5 per cent, and 9.7 per cent of the respective world averages,[11] and that China's old model of industrialization characterized by high investment and high consumption of energy resources have impelled China to search for energy resources overseas, including in Southeast Asia. The prevailing views in Southeast Asia are that, "in the eagerness to deploy Chinese capital and expertise for rapid completion of resource extraction, transportation and power-generation projects, Chinese SOEs (state owned enterprises) have been given wide leeway in disregarding environmental standards and the interests of local people affected by these projects".[12] Many Southeast Asian observers are concerned that China will replicate the sort of "neo-mercantilist" strategies that Japan adopted during its high-growth phase in the 1970s, a possibility that is reinforced by the prominence of "state capitalism" in a number of rising powers.[13]

Moreover, according to Acharya, three factors have played a major role in shaping ASEAN's concerns over China's rise: (1) China's involvement in the Spratly Islands disputes; (2) China's military expansion; and (3) suspicion over an increased "overseas Chinese presence" and its implications for interethnic relations among some ASEAN countries.[14] While most ASEAN leaders do not see China as an immediate threat and see recent developments as promising and reassuring, there is still concern about how deep recent changes might run.

Here, power differences and asymmetry, made significant by both their recent and not so-recent relations, pose an important obstacle to China's ability to convince ASEAN countries that its intentions are benign. Thus, "even given China's and ASEAN's common sense of vulnerability and common grievances against larger Western powers, China remains a major power in the eyes of ASEAN".[15] This suggests that ASEAN governments continue to view China's foreign policy with some measure of mistrust and suspicion in regards to the stability of the region, especially in the South China Sea.[16] This is especially so in light of Beijing's recent growing "assertiveness" in terms of its energy resource exploration, maritime claims, and frequent military activities in the South China Sea. In this process, it is natural to see that ASEAN subtly began to shift from its "ASEAN way" of quiet bilateral conflict settlement to the search for a multilateral solution, as China was too big and too powerful to face bilaterally.

Hence, the rise of China's power and its influence in the world has had dual impacts on Southeast Asia. On the one hand it does create business and economic opportunities for Southeast Asian countries as it has become nearly every East Asian country's largest trading partner and increasingly important investor. On the other hand, it also results in rising concerns of ASEAN countries in terms of resource exploitation, regional security, and possible conflicts in the South China Sea. So far, the main response of the Southeast Asian countries to China's rise has been to seek regional economic integration and accelerate the establishment of an ASEAN Community, and also encourage outside big powers including the United States, Japan, and India to remain involved in the region, especially in the South China Sea.

ASEAN'S RISE IN THE GLOBAL ECONOMY

ASEAN was established on 8 August 1967 with the signing of the ASEAN Declaration by Thailand, the Philippines, Indonesia, Singapore, and Malaysia.[17] ASEAN is the most advanced institution of regional cooperation in Asia and one of its oldest. At first, its goals were mainly political in nature. In particular, it sought to promote peace in what was at that time a volatile region. While these diplomatic initiatives did not promote economic integration directly, the peace and security that followed paved the way for economic growth and development throughout Southeast Asia,[18] making it another emerging economy in the world.

In the late 1980s and early 1990s, ASEAN took steps to develop a free trade area. This was in marked contrast to the lackadaisical attitude.[19] That attitude, as an Australian professor, Stubbs points out, had to do with a concern for sovereignty as well as domestic economic conditions in the member states. Until the late 1980s, the most populous ASEAN countries — Indonesia, the Philippines, Thailand and Malaysia — had remained heavily reliant on raw material exports and import-substitution strategies. Indonesia's oil boom of the 1970s discouraged export promotion strategies. In Malaysia, the advent of the New Economic Policy (NEP) (aimed at giving indigenous Malays a greater share of the national wealth) resulted in massive government intervention, especially in creating import-substituting heavy industries. These conditions lessened the urgency of intra-regional trade liberalization, more commonly associated with economies geared towards export promotion.[20] Moreover, the level of intra-ASEAN trade had remained fairly low due to colonial linkages and the impact of the Vietnam War. As the 1990s approached, ASEAN members' trade with the United

States, Western Europe, and Japan was considerably higher than with each other.[21]

A number of developments led to a shift in the attitude of ASEAN states toward significant economic cooperation initiatives, including rising protectionism in the United States, the economic recession in the ASEAN countries in the early 1980s. In addition to this, as Severino points out, this was also partially in response to China's economic rise.[22] Some Southeast Asian and Western observers do worry about the region becoming a vassal of China. They foresee the emergence of a Chinese sphere of influence in Southeast Asia, or a Chinese Monroe Doctrine.[23] Southeast Asians are also concerned about economic marginalization. The single most important concern has been investment diversion to China.

ASEAN's first major initiative was AFTA (ASEAN Free Trade Agreement), which was concluded at the Singapore Summit in 1992. The treaty called for a reduction of tariffs on intra-ASEAN trade in a fifteen-year period and the abolition of qualitative restrictions and other non-tariff barriers to trade in time-specific ways. An ASEAN agreement in January 2003 provided for the eventual elimination of tariffs on intra-ASEAN trade.[24] The ASEAN governments agreed on a common set of tariff nomenclatures both to make it easier for their countries' companies to trade with one another and for themselves to strike trade deals as a group with other countries.

ASEAN has also made important strides in the area of investment cooperation (in the form of ASEAN "one-stop investment centres" and the ASEAN Investment Area), and trade facilitation (customs cooperation). The idea was to reduce transaction costs associated with intra-regional economic interaction, making Southeast Asia, with its vast combined consumer base, more attractive to foreign direct investment again in the face of China's rise. ASEAN has reached this goal to some extent. For example, from 2007–12, China's FDI (foreign direct investments) inflows increased from US$83.5 billion to US$121.1 billion, while ASEAN's FDI inflows increased from US$85.6 billion to US$111.3 billion, becoming the main FDI destinations in developing economies.[25] By 1 January 2010, AFTA was duly established among ten ASEAN countries. By the end of 2010, tariffs on 99.11 per cent of products among ASEAN-6 were eliminated, and tariffs on 98.86 per cent of products of CLMV (Cambodia, Lao PDR, Myanmar and Vietnam) countries were reduced to below 5 per cent.[26]

In the process of AFTA, ASEAN began working to establish ASEAN economic, security, and social-cultural communities. In November 2002, the ASEAN leaders agreed, at the Bali ASEAN Summit in October 2003, to create an "ASEAN Economic Community" (ACE) by 2020. In 2007 "Cebu Declaration", the ASEAN leaders pushed up the deadline to 2015. The goals of ACE are: (1) to create a single market and production base, including to facilitate free flow of goods, services, investment, capital, and skilled labour; (2) to form a competitive economic region, including to introduce competition policy, establish consumer protection measures, fully implement ASEAN intellectual property rights policy, and promote infrastructure development; (3) to promote equitable economic development, including to accelerate the pace of SME (small and medium-sized enterprises) development, and enhance the overall economic growth and development of ASEAN as a region; (4) to be integrated into the global economy. According to the ACE blueprint, "ASEAN shall work towards maintaining 'ASEAN Centrality' in its external economic relations, including, but not limited to, its negotiations for FTAs and comprehensive economic partnership (CEPs) agreements."[27]

With the rapid economic rise of ASEAN and the accelerated regional economic integration, ASEAN's position in Asia-Pacific region continues to improve. According to statistics, the total GDP of ASEAN countries in 1996 was US$725.5 billion, decreased to US$472.6 billion in 1998 because of the Asian financial crisis, recovered to US$796.5 billion in 2004, and further increased to pass US$2 trillion in 2011.[28] By 2012, ASEAN has a total land area of 4.43 million square kilometres, total population of 617 million, aggregate GDP of US$2.3 trillion, total import and export value of US2.5 trillion, representing the fifth largest economy in the world (only after the European Union, United States, China, and Japan).[29] ASEAN is, as a result, likely to play a far more significant role in the global economy as the twenty-first century unfolds.

ASEAN is a bulwark of regional stability and increasing prosperity in Southeast Asia and a pivotal element in the geopolitics of the whole Asian region.[30] It has become a central feature of Asia-Pacific regional architecture, and to the extent that it will deepen its trade links, will play a larger role in future Asia-Pacific regional cooperation. ASEAN member states, Brunei, Malaysia, Singapore, and Vietnam, are negotiating a Trans-Pacific Partnership (TPP) with the United States, Japan, Canada, Chile, Peru, Mexico, Australia, and New Zealand. Negotiations are also continuing to set up the Regional Comprehensive Economic Partnership (RCEP). This is designed to link ASEAN's ten member countries with New Zealand, Australia, South Korea, India, Japan, and China into one regional FTA. Together, these countries account for almost half the global population and about a third of global output, and the share of the intra-regional trade reached over 40 per cent, making the region the world's new economic centre of gravity.

NEW GLOBAL ENERGY ORDER

The rise of China, ASEAN, and other Asian countries like India economically has prompted East Asia to become the world's new economic centre of gravity and inevitably created a great impact on the world energy market, and will hence accelerate the changes in world energy system, accelerating an energy shift from west to east.

Rising oil and gas imports are trending across Asia, in particular as China surpasses the United States to become the world's largest importer of crude oil and ASEAN as a whole, changes from energy resource exporter to energy importer. Japan's already high dependence on imported oil and gas supplies has skyrocketed after the country's nuclear reactors were taken offline following the Fukushima nuclear disaster. South Korea revised downward its plans for expanding its own nuclear power sector, which will have a negative impact on its plans to diversify energy supply. Meanwhile, growing demand in ASEAN will increasingly impact global energy markets. In a recent report on Southeast Asia, the International Energy Agency (IEA) predicts that ASEAN will become the world's fourth-largest oil importer by 2035 as its net import-dependence will almost double to 75 per cent and net imports rise from 1.9 million barrels per day (mb/d) to just over 5 mb/d.[31] Rising Asian demand, spurred by growth in China and ASEAN countries, has been a critical driver of increases in global energy demand.

In the past years, with the surging Asia energy demand and soaring North American shale gas, the world energy system has experienced dramatic changes. The first is energy demand (oil and natural gas) shifts from west to east. The era of growing demand for oil and other fossil fuels in the industrialized countries is over; most of the future growth in demand will come from the emerging countries in Asia. According to IEA, the share of oil demand from the Organisation for Economic Co-operation and Development (OECD) countries in total world oil demand had declined from 65 per cent in 1980 to 56 per cent in 2006, and is predicted to further decline to 46 per cent in 2030; while that from Asian developing countries (excluding Middle Eastern countries) increased from 6.7 per cent in 1980 to 18.0 per cent in 2006, and is predicted to further increase to 27.5 per cent by 2030.[32]

In terms of natural gas, according to IEA, the share of its demand from the OECD countries in total world gas demand had declined from 63 per cent in 1980 to 51 per cent in 2005, and is predicted to further decline to 42 per cent in 2030; while that from Asian developing countries (excluding Middle Eastern countries) had increased from 2.4 per cent in 1980 to 9.2 per cent in 2005, and is predicted to further increase to 14.8 per cent by 2030.[33]

Rising energy consumption (especially coal and oil) in developing Asia is contributing to higher CO_2 (carbon dioxide). According to IEA, the share of energy-related emissions comes from OECD countries will decrease from 48 per cent in 2005 to 36 per cent in 2030, while that from the Asian developing countries will increase from 28.6 per cent to 42.5 per cent (excluding Middle East). Most of the increase in energy-related CO_2 emissions comes from China, India, and some ASEAN countries which together account for over 60 per cent of the increase in emissions between 2005 and 2030. China is by far the biggest single contributor to incremental emissions between 2005 and 2030.

The second structural change is the geographic location of the energy sector. While the remaining oil and gas reserves are concentrated in the Middle East and Persian Gulf, Africa, and Central Asia and Russia, in the other hemisphere the United States is emerging as a global energy producing giant in its own right. According to IEA World Energy Outlook 2013, as a result of tight oil and shale gas development, the United States, who has been the largest producer of natural gas since 2012, is expected to become the world's largest oil (crude, unconventional plus natural gas liquids) producer in 2015, and is expected to remain so through early 2030.[34] U.S. gas and oil production increase along with its slower demand growth has decreased the country's need for imports. As a result, traditional U.S. suppliers are increasingly servicing other markets, suggesting that the Indo-Pacific region will become increasingly reliant on the Middle East for oil.

At the same time, Russia is increasingly shifting its focus of energy exports from Europe to East Asia, and China is expected to become Russia's biggest export market of oil and gas soon. These dynamic changes have great impacts on the producer-supplier relationships and the energy trade flows.

The third shift in world energy system is the energy structure change. The potential ability of natural gas to serve as a substitute for coal and oil is important. Japan's greater reliance on liquefied natural gas (LNG) to offset the deficit in nuclear power has reshaped outlooks for LNG markets. More importantly, the argument is that gas provides an obvious transition fuel to a lower carbon economy, displacing coal, especially after the shale gas revolution increases supply and keeps prices low. For example, in 2011 emissions of CO_2 in the United States dropped to their lowest level since 1995. In the European Union and United States and other parts of the world, relatively low gas prices and the rising carbon price meant that it was more expensive to generate electricity from coal than from gas.[35]

The importance of gas and the increase of its production were partly due to the difficult development of clean energy and the result of fundamental technological breakthroughs in U.S. gas production. Technological advances, such as horizontal drilling (which eases access to layers of oil or gas) and hydraulic fracturing (which uses water pressure to release gas from hard rocks) were employed to make unconventional gas resources, such as tight gas, shale gas, and coal-bed methane, accessible on a large scale.[36] Doubtless, the shale gas revolution will lead to a great change in global energy development.

Another driving factor for gas use is the growing concern about carbon emissions and a growing awareness that fossil fuels cannot immediately be replaced with carbon-free alternatives. Gas burns much more cleanly than coal and oil. Producing one kilowatt-hour of electricity with natural gas emits a little more than half the amount of carbon that producing the same amount of energy with coal does.

The third factor is the difficulties for developing other alternative clean energies: it is technically and economically not yet feasible to make renewable energy such as solar power, wind power, and biofuels the main source of energy; the nuclear power crisis that happened in Japan in the early 2011 has set back the nuclear power development in the world and Southeast Asia. Many countries like Germany, France, United States, and Indonesia have announced that they will abandon or postpone developing new nuclear power projects.

The above changes in the world energy system have strategic implications for Asian countries, especially for China and ASEAN countries, and pose some critical questions for us to consider. Firstly, as China's energy demand rises and its broader energy security strategy has been to pursue supply diversity, to find more and develop more offshore oil and gas resources, China has the intention to invest more on natural gas sectors in some ASEAN countries, such as Indonesia, Myanmar, and Malaysia. Can these countries work more closely in the gas field, including gas pipeline, other energy-related infrastructures building and gas exploration? In the background of China's rise and its growing influence in Southeast Asia, will China's intention for energy resource cooperation be looked as a threat (source of conflicts) or opportunity to its neighbouring countries? Given that there is great potential for energy cooperation between China and ASEAN countries, can the energy sector serve as a positive factor in deepening China–ASEAN bilateral relations, or vice-versa?

Secondly, as the world energy shifts east, and as the Greater Indian Ocean and the South China Sea become the world's energy interstate, maritime tensions are rising in the South China Sea and in the adjacent East China Sea. The territorial tensions in those claimed waters are not only driven by potential energy reserves and fish stocks, but also by the fact that these sea lanes and choke points are of growing geopolitical importance because of the changing world economic and energy market. Then, will geopolitical tensions over energy resources spark conflicts in the region, such as in the South China Sea? Can these tensions be reduced or conflicts be avoided through "joint development"?

Thirdly, since the United States, Japan, and India are important players in Southeast Asia, does the shifting geopolitics of energy give these big powers a new strategic tool in an intensifying rivalry with China? Or does the changing geopolitics of energy create more areas of shared interests and opportunities for cooperation between these big powers, to balance rather than increase tensions in Southeast Asia?

DEBATES ON ENERGY RESOURCE COOPERATION AND COMPETITION

Competition for scarce natural resources has been an important determining factor in human development. In history, tribes of hunter-gatherers fought over land and the flora and fauna that surrounded them, and early agricultural societies that existed along rivers fought deadly conflicts over getting their share of the water. Kingdoms, large and small, traded or battled for iron, gold, and other metals, as well as precious stones. The beginning of the Industrial Revolution in Western Europe and the input materials it required were major reasons for the expansion of colonialism.[37]

In current world society, the core issue and main concerns of many related countries are whether competition for strategic resources will lead to conflicts between nations, whether energy cooperation can strengthen bilateral relations, and what possible impacts a big country's energy quest strategy will create.

Will energy competition lead to possible conflicts?

(1) Competition for Strategic Resources May Not Necessarily Lead to Conflicts — Resource Optimism Arguments

Scholars who hold this view believe that strategic resources can always be replaced by other resources (such as renewable energy resources). Therefore, competition for strategic resources will not lead to or exacerbate conflicts. Leonardo Maugeri stands in sharp opposition to the "oil doomsday prediction", being optimistic about the prospects for oil supply.[38] He believes that due to the rising oil prices, the use of oil has been replaced by other energy products. For example, the oil had once been used extensively for power generation, but now there are few oil-fired power plants. Oil power generation has largely been replaced by natural gas and uranium.

Maugeri also referred to the internal oil substitute. The oil we use most today often refers to the conventional oil resources which is only a part of oil resources. In addition to these conventional oil resources, there are a lot of unconventional oil resources on the planet, including heavy oil, shale oil, tar sand, etc. When oil prices rise to be favourable enough for profitable exploitation of unconventional oil, then these unconventional oil will be able to substitute more oil.[39]

Moreover, technology and machinery used for oil is in rapid upgrading due to technology innovation and development. This is also resource substitute. In the transportation sector, for example, although the dominant position of oil cannot be shaken, the engines driven by petroleum products (gasoline and diesel) are in constantly upgrading and replacement. The new more fuel-efficient internal combustion engines continuously replace old inefficient internal combustion engines.

Maugeri did not explicitly deny the theory that oil resources are limited, but he put forward that due to a number of factors, the day of oil depletion will be indefinitely postponed although occasional oil shortage cannot be completely avoided. Hence, he actually believes that oil resources are unlimited. Since oil resources are unlimited, then oil relations between countries should be non-zero sum. "Nothing should let us compete brutally for ensuring future oil supply in the face of extreme shortages and regional chaos."[40]

(2) Energy Resources May Lead to Possible Conflicts — Resource Pessimism Arguments

The Geography of Conflict is a branch theory of resource pessimism. It was raised by Michael T. Klare.[41] This theory mainly describes the relationship between natural resource endowments and conflicts between nations, with strong pessimistic colours. Klare proposed that as populations increase and economic activity expands in many parts of the world, the appetite for vital materials will only swell more quickly than nature can accommodate, resulting in resource supply not being able to keep up with the demand. Technologies that introduce alternative materials and production techniques will help overcome some of these scarcities, but cannot completely reverse demographic and environmental pressures, and some countries and regions will be unable to afford the higher costs of alternative technologies.

In these circumstances, global supply and demand will become increasingly unbalanced. Klare believes that as shortages of critical resources rise in frequency and severity, the competition for access to the remaining supplies of these commodities will only grow more intense in years to come.[42] Hence, it can be believed that the potential for regional conflict grows in light of the fact that China, India, and some ASEAN countries such as the Philippines, Thailand, and Vietnam have growing dependence on foreign energy resources.

Another argument on possible conflicts is based on the views that a nation may have "security dilemma" on another nation's certain military behaviour to secure its energy shipping routes, and a country's military expansion will inevitably lead to conflicts. Realists believe mistrust between potential enemies could lead to misperception, and misperceived offensive threats can lead to countermeasures in kind, resulting in an unnecessary spiral of tensions and the danger of arms race and war.

For example, as China becomes a greater economic power, it will become increasingly dependent on shipping routes for its imports of energy resources, other inputs and goods. China's vulnerability to having these imports choked off is acute and possible. This implies the need to develop a blue-water navy to ensure that China's economy cannot be strangled by a maritime blockade.

But seeing from a realist prism, what China considers a defensive imperative could be easily perceived as aggressive and expansionist by its neighbours and the United States. Similarly, what looks like a defensive imperative to the United States and its Asian allies — building further military capacity in the region to manage China's rise and military expansion — could be perceived easily as well by China as an aggressive attempt to contain it.[43] Hence, it is not surprising to see that "should China seek to protect its sea lanes, the U.S. Department of Defense would see this as potentially challenging the U.S. navy's accustomed role in protecting international sea lanes and as China being capable of involvement in territorial or resource wars".[44] Thus energy is arguably a driver of one of the most complex tensions in modern politics: the naval arms race between the United States and China, centred for now in the East and South China Seas.[45]

(3) Dynamic Nature of the Relationship Between Humanity and Resources

The differences between the above resource pessimism and optimism can be summed up in the following aspects. On the natural resources, resource pessimism believes that resources are limited, while resource optimism thinks that resources are unlimited. On human resource relationship, resource pessimists think that the human resource relationship is zero-sum, while the optimists consider it to be non-zero sum. Resource pessimists think that the lack of resources would inevitably lead to competition, conflict, or even war for resources; while optimists believe that human competition for resources and the outbreak of war are not common phenomena in human history.

Here the argument point of resource pessimism is that since resources are limited, they are bound to run out eventually. When resources are nearly depleted, there will be tensions between supply and demand, then the competition for resources will lead to outbreak of conflicts and even war.

Whereas the argument point of resource optimism is that resources are unlimited. Some resources seem to be limited, this is because the human's ability to exploit and utilize these resources is low, or because of a temporary lack of alternative means. Once the tensions between supply and demand rise, people will inevitably try to improve the efficiency of resource exploitation and utilization, and find alternative means. As a result, the tensions between supply and demand can be released and will disappear eventually. Therefore, conflicts or wars for resource competition will not happen.

These two views or arguments on human resource relations are convincing for their own logics. However, they have a common flaw. Their discussions are based on an assumption which can only be consistent in some cases. For example, resource pessimists assume that natural resources are non-replaceable, or assume that even if there exists alternative resources, the substitution effect is very small. On the other hand, resource optimists believe that alternative means of resources will always appear and play an effective role of substitution. If there are always plenty of alternative resources, the possibility of the occurrence of resource conflicts will be very small. Since the basic assumptions of both resource pessimism and optimism are fixed and static, they are static resource theory.

Clearly, for resource pessimism and optimism, the question of whether resources are replaceable or not is certain and fixed, therefore, whether conflicts will happen is certain as well. However, for most resources whether they are replaceable or not is actually an uncertain question. In addition to technical factors, there are two major factors that can change the possibility of whether resources can be replaceable or not.

The first factor is the economic pressure created by resource shortage. Under such economic pressure, people will try to make alternative means of resources, including the development of resource-saving technologies, to find alternative resources, and to exploit the same resources in a wider range, so as to increase the supply of resources and reduce resource dependency. The second factor is the competition and confrontation caused by a shortage of resources. Because energies and funds are used for resource competition, confrontation or even war, this would largely delay the development of alternative means, making resource shortages even more severe.

Hence, the relationship between states in the competition for natural resources may change the above two factors. If serious competition or conflicts happen because of the shortage of resources, then, there is no guarantee that enough technologies and human capital will be used for developing alternative resources.

Therefore, the alternative means of scarce resources is related to the status of resource relations between related states in their quest for natural resources. If nations ignore the existence of alternative means of resources, see their resource relationship as a zero-sum relationship, and take confrontation actions based on this concept, then the possibility of resource conflicts is likely to be great. Instead, if states attach importance to the development of alternative means of resources, see their resource relationship as a non-zero-sum relationship, and take cooperative actions based on this concept, then the likelihood of resource conflicts will be largely reduced.

Can energy cooperation strengthen bilateral relations?

(1) Energy Cooperation Can Cement Bilateral Relations

Increased energy resource trade and project investment provide related countries with a more solid foundation for future partnerships. Based on this view, the Ministry of foreign Affairs of China argues that economic interdependence is an integral component of regional security, and that China will push for greater regional economic cooperation to ensure the development of such security.[46] Many international scholars also argue that energy cooperation can deepen regional integration and thus strengthen bilateral relations.

Saleem H. Ali argues, for example, that due to the permanence of their infrastructures, such as oil and gas pipelines and other energy projects, in strengthening interstate relations, these cooperative projects are likely to have a more lasting impact and create greater incentives for cooperation over time.[47] For the supply countries, pipelines and related infrastructure projects can provide much-needed employment and revenue for the host countries, in the process quelling some of the domestic resentment that fuels extremism.

More importantly, Ali also believes that related countries can utilize pipeline construction project as both an engine of cooperation and a tool of diplomacy. Pipelines open up regions for development and have spillover effects into downstream industries such as factories, chemical and fertilizer facilities, and refineries that have incentives to locate themselves close to sources of natural gas supply. Hence, "rather than being a source of conflict, energy has the capacity to become an integrative force, creating a large sense of shared interests and stakes in cooperation".[48]

(2) Energy Cooperation and Regional Economic Integration May Not Necessarily be a Source of Stability

Michael Yahuda of the London School of Economics argues that economic interdependence between China and the small states in Asia has not led to greater stability, as historical and contemporary political issues remain the defining characteristics of such relations.[49] Chinese investment in Southeast Asia brings the countries closer, but it also reminds politicians of "new colonialism" and mercantilist instincts, raising energy resource nationalism among host countries and generating more sources of friction. Populist cries for governments to exercise sovereign power decisively make it harder for state officials to work toward a mutually acceptable compromise.

While Paul Stevens believes that cross-border pipelines and related projects can generate conflicts and local resentments, as parties with different interests and motivations are involved, and land use cannot be compensated properly. "This invites disagreement because of the benefits to be shared and mechanisms exist to encourage both parties to seek a greater share."[50] Jia Qingguo of Peking University also argues that economic interdependence cannot ensure good relations between China and the small states if the latter are uncertain about China's foreign policy intentions.[51] Zhao Suisheng noticed that China's tactic of putting aside areas of disagreement in favour of creating a stable environment for economic development are limited to areas where China's vital strategic interests are not threatened.[52] Bluntly put, Beijing's long-term strategic intentions might inspire deep anxieties and concerns of the local people and governments.

DIFFERENT VIEWS ON CHINA'S ENERGY QUEST STRATEGY

Most Western scholars hold that the "going out" strategy cannot fundamentally solve a country's energy security.[53] "The main means of addressing China's energy security concerns has been to rely on markets, which make it easier to diversify supply and demand, substitute fuels, and make the most of the gains in efficiency brought on by technological change".[54] Energy security is a global problem that requires a global solution, national energy security depends on international energy security.[55] They believe, actually, that the crux of China's energy problem is the lack of rational and efficient energy management system.[56] It needs China to further reform its energy system, for example, allowing more private companies to participate in energy sectors.

Meanwhile, the "going out" strategies for energy resources have different impacts. China's oil import dependence has put energy security high on China's foreign policy agenda. It pursues political relations with oil and gas producing countries, looking for bilateral agreements for future oil and gas supplies; and, through its NOCs (National Oil Companies), it has engaged in mutual investment relations in the host countries, often seeking to construct related infrastructure and energy projects. While these actions create positive benefits for local people, however, they are regarded as security threats or source of conflicts by many U.S. analysts and politicians.[57] For example, it is believed that China's energy cooperation with Sudan, Iran, and Myanmar could lead to increased regional and local conflicts. "China's close relationships with oil-producing nations in the Gulf region, particularly those non-U.S.-friendly, have raised American eyebrows."[58] Thus China's strategic pressure has been increasing.[59]

Indeed, China's state-centred approach toward energy security has led to a mercantilist strategy to bolster energy supplies by gaining direct control of oil and gas fields and supply routes. This strategy has produced a mixed result in its foreign relations. On one hand, it has brought an opportunity to enhance cooperation with its neighbours. On the other, this is possible to destroy market order, erode confidence in fair access to future supplies and aggravate strategic distrust.[60] This strategy has contributed to mounting distrust and concerns in local communities (such as in Myanmar and Indonesia), even though the level of direct state intervention varies.

Zhu Feng, a professor of Peking University, recognized that Beijing had been overly focused on building relations with Myanmar's government and ignored the feelings and interests of local people. Heinrich Kreft also warned that "the results of China's resource diplomacy are being watched with growing unease, especially in Asia", and he believed that "there is a danger that China's neo-mercantilist strategy to bolster energy security by gaining direct control both of oil and gas fields and supply routes could result in escalating tensions in an already volatile region that lacks regional institutions for conflict resolution."[61] Hence, these scholars hold that China's request for energy resources may become the spark for regional and international instability.

However, Llewelyn Hughes believed that "locking up oil does not matter", because "even we allow that energy policies in the rising Asia-Pacific powers are government-led rather than firm-led, and that they are designed to enhance energy security, we nevertheless need not fear that this leads to zero-sum dynamics, as there are ample commercial opportunities available to NOCs from Asia-Pacific countries."[62] Moreover, Jeffrey D. Wilson held that resource security strategies are not without historical precedent in Northeast Asia. They argued that during the 1970s and 1980s, the Japanese as well as the Korean government offered financial and diplomatic assistance to their industrial corporations to sponsor the development of new mining firms in Latin America and Southeast Asia.[63] Thus China's "going out" strategy would create more positive impacts than negative impacts.

Notes

1. Alice D. Ba, "China and ASEAN: Renavigating for a 21st Century Asia", *Asian Survey* 43, no. 4 (2003): 622–47.
2. Ibid.
3. Joseph Y.S. Cheng, "China's ASEAN Policy in the 1990s: Pushing for Regional Multipolarity", *Contemporary Southeast Asia* 21, no. 1 (1999): 176–205.
4. Alice D. Ba, "China and ASEAN: Renavigating for a 21st Century Asia", *Asian Survey* 43, no. 4 (2003): 622–47.
5. Alice D. Ba, "Who's Socializing Whom? Complex Engagement in Sino–ASEAN Relations", in *Theorizing Southeast Asian Relations*, edited by Amitav Acharya and Richard Stubbs (Abingdon: Routledge, 2009).
6. Chinese scholars consider the financial crisis a watershed event regarding ASEAN countries' perceptions of China. The year 1997 is the landmark to indicate the rise of China's soft power in Southeast Asia. See Chen Xiansi, "On China's soft power in Southeast Asia", *China's Foreign Affairs* 5 (2007): 32–33.
7. Zheng Bijian, "China's 'Peaceful Rise' to Great-Power Status", *Foreign Affairs* (Sep/Oct 2005).
8. On 7 September 2013 in Kasakhstan, President Xi announced a new policy of building a "community of shared interest" with Central Asia, and on 3 October the same year, Xi proposed to build a "community of shared destiny" with Southeast Asia.
9. *China Customs Statistics Monthly*, December 2013.
10. Ministry of Commerce of China, *2012 Statistical Bulletin of China's Outward FDI*, p. 23.
11. Zheng Bijian, "China's 'Peaceful Rise' to Great-Power Status", *Foreign Affairs* (Sep/Oct 2005).
12. John Lee, "Myanmar Pivots Awkwardly Away from China", *ISEAS Perspective*, 12 December 2013.
13. Mark Beeson, Mills Soko, and Wang Yong, "The New Resource Politics: Can Australia and South Africa Accommodate China?" *International Affairs* 87, no. 6 (2011), p. 366.
14. Amitav Acharya, *Regionalism and Multilateralism: Essays on Cooperative Security in Asia-Pacific* (Singapore: Times Academic Press, 2002), p. 200.
15. Alice D. Ba, "Who's Socializing Whom? Complex Engagement in Sino–ASEAN Relations", in *Theorizing Southeast Asian Relations*, edited by Amitav Acharya and Richard Stubbs (Abingdon: Routledge, 2009).
16. Denny Roy, "Southeast Asia and China: Balancing or Bandwagoning", *Contemporary Southeast Asia* 27, no. 2 (2005): 305–22.
17. Rodolfo C. Severino, "Will There Be A New ASEAN in the 21st Century?", *Asia Europe Journal* 2, no. 2 (2004): 179–84.
18. Michael G. Plummer, *ASEAN Economic Integration: Trade, Foreign Direct Investment, and Finance* (Singapore: World Scientific, 2009), p. 22.

19. Amitav Acharya, *The Making of Southeast Asia* (Singapore: Institute of Southeast Asian Studies, 2012), p. 224.
20. Ibid.
21. Richard Stubbs, "Signing on to Liberalization: FTA and the Politics of Regional Economic Cooperation", *The Pacific Review* 13, no. 2 (2000): 297–318.
22. Rodolfo C. Severino, "The Rise of Chinese Power and the Impact on Southeast Asia", *ISEAS Perspective*, 27 May 2013.
23. Amitav Acharya, *The Making of Southeast Asia* (Singapore: Institute of Southeast Asian Studies, 2012), p. 269.
24. Rodolfo C. Severino, "The Rise of Chinese Power and the Impact on Southeast Asia", *ISEAS Perspective*, 27 May 2013.
25. UNCTAD, *World Investment Report 2013: Global Value Chains: Investment and Trade for Development*, p. 214.
26. Wang Qin, "dongmeng quyu jingji yitihua de geju jiqi yingxiang" [The Pattern of ASEAN Economic Integration and Its Impact], *Southeast Asian Affairs* 156, no. 4 (2013).
27. ASEAN Secretariat, *ASEAN Economic Community Blueprint*, pp. 6–25, <http://www.asean.org/archive/5187-10.pdf> (accessed 30 March 2015).
28. Ibid.
29. ASEAN Secretariat, *ASEAN Community in Figures ACIF 2012*, 2013.
30. Peter Drysdale, "Economic Community Key to ASEAN's Centrality", *East Asian Forum*, 12 May 2014.
31. IEA, *Southeast Asia Energy Outlook 2013*, pp. 11–12.
32. IEA, *World Energy Outlook* 2007, p. 80.
33. Ibid., p. 85.
34. IEA, *World Energy Outlook 2013*, pp. 73–76.
35. Christof Ruhl, "Global Energy After the Crisis: Prospects and Priorities", *Foreign Affairs*, March/April 2010.
36. Ibid.
37. Toufiq Siddiqi, "China and India: More Cooperation than Competition in Energy and Climate Change", *Journal of International Affairs* 64, no. 2 (Spring/Summer 2011).
38. Leonardo Maugeri, *The Age of Oil: The Methology, History, and Future of the World's Most Controversial Resource* (Westport Connecticut: Praeger Publishers, 2006).
39. Ibid., p. 214.
40. Ibid., p. xxi.
41. Michael T. Klare, "The New Geography of Conflict", *Foreign Affairs* 80, no. 3 (2001): 49–61.
42. Ibid.
43. Nouriel Roubini, "Keeping China's Rise Peaceful is Our Biggest Geopolitical Challenge", *The Guardian*, 30 April 2014, <http://www.theguardian.com/business/2014/apr/30/project-syndicate-china-peaceful-rise-biggest-geopolitical-challenge?> (accessed 30 March 2015).

44. Stuart Harris, "Global and Regional Orders and the Changing Geopolitics of Energy", *Australian Journal of International Affairs* 64, no. 2 (April 2010).
45. Bruce Jones et al., "Fueling a New Order: The New Geopolitical and Security Consequences of Energy", Project on International Order and Strategy at Brookings, <http://cic.nyu.edu/sites/default/files/publication_april2014_14_geopolitical_security_energy_jones_steven.pdf> (accessed 15 May 2014).
46. Ministry of Foreign Affairs of the People's Republic of China, "Zhongguo guanyu xin anquan guan de lichang wenjian" [Document on China's Position Regarding Its New Security Concept], 31 July 2002, <http://www.fmprc.gov.cn/chn/gxh/zlb/zcwl/t4549.htm> (accessed 15 May 2014).
47. Saleem H. Ali, "The Strategic Logic of Pipelines: Toward 'Rational Regionalism'", The Brookings Institution paper, 2010.
48. Benjamin K. Sovacool, "Energy Policy and Cooperation in Southeast Asia: The History, Challenges, and Implications of the Trans-ASEAN Gas Pipeline (TAGP) Network", *Energy Policy* 37 (2009).
49. Michael Yahuda, "Chinese Dilemmas in Thinking About Regional Security Architecture", *The Pacific Review* 16, no. 2 (2003).
50. Paul Stevens, "Oil and Gas Pipelines: Prospects and Problems", *NBR Special Report #23*, September 2010.
51. A speech by Peking University's vice director of the School of International Relations, Professor Jia Qingguo, *Chahare xuehui*, 29 December 2011.
52. Zhao Suisheng, "China's Global Search for Energy Security: Cooperation and Competition in the Asia-Pacific", *Asia-Pacific Journal* 17, no. 55 (2008): 207–27.
53. Erica Downs, "China's Quest for Overseas Oil", *Far East Economic Review*, September 2007.
54. Christof Ruhl, "Global Energy After the Crisis: Prospects and Priorities", *Foreign Affairs* 89, no. 2 (March/April 2010).
55. Erica Downs, "National Energy Security Depends on International Energy Security", The Brookings Institute website, <http://www.brookings.edu/opinions/2006/0317china_downs.aspx?p=1> (accessed 15 August 2015).
56. Zha Daojiong, "cong guoji guanxi jiaodu kan zhongguo de nengyuan anquan" [China's Energy Security from the Perspective of International Relations], *Guoji Jingji Pinglun* [Review of International Economy], no. 11 (2005).
57. Stuart Harris, "Global and Regional Orders and the Changing Geopolitics of Energy", *Australian Journal of International Affairs* 64, no. 2 (April 2010).
58. Pak K. Lee, "China's Quest for Oil Security: Oil (Wars) in the Pipelines?" *The Pacific Review* 18, no. 2 (2005).
59. Zha Daojiong, "cong guoji guanxi jiaodu kan zhongguo de nengyuan anquan" [China's Energy Security from the Perspective of International Relations], *Guoji Jingji Pinglun* [Review of International Economy], no. 11 (2005).

60. Gabe Collins et al., "Asia's Rising Energy and Resource Nationalism: Implications for the U.S., China, and the Asia-Pacific Region", National Bureau of Asian Research, NBR Report, September 2011.
61. Heinrich Kreft, "China's Quest for Energy", *Hoover Policy Review*, no. 139, 1 October 2006, <http://www.hoover.org/publications/policy-review/article/7941> (accessed 15 August 2015).
62. Llewelyn Hughes, "Resource Nationalism in Asia-Pacific: Why Does It Matter?", *NBR Special Report #31*, September 2011.
63. Jeffrey D. Wilson, "Northeast Asian Resource Security Strategies and International Resource Politics in Asia", *Asian Studies Review* 38, no. 1 (2014): 15–35.

2

ECONOMIC GROWTH AND ENERGY SECURITY

China and most ASEAN countries are characterized by high economic growth and rapid increase in energy resource consumption. Although these countries are trying hard to find appropriate energy alternatives, increase energy efficiency, and diversify energy resources, they are becoming increasingly dependent on imported oil and gas, and the unstable Middle East and Africa are set to remain their predominant sources of oil and gas. Thus high dependence on fossil fuels poses several security concerns and challenges, including a widening gap between supply and demand, transportation of oil and gas resources on vulnerable sea routes, and energy-related CO_2 emissions. This chapter mainly analyses these three dimensions in the context of regional and global perspectives.

ECONOMIC GROWTH AND ENERGY DEMAND

Energy Demand Drivers

The demand for energy resources in Asia is predominately driven by two factors. The first is "industrial-led" demand related to economic growth, a structural shift from non-mechanized forms of manufacturing and production to more energy-intensive ones, especially for commodities such as iron and steel, cement and glass, paper and pulp, basic chemicals, and non-ferrous metals. The second is "consumption-led" demand as population grows and living standard rises (urbanization), bringing more energy-intensive lifestyles that revolve around automobiles, air conditioning, and disposable goods.

Economic Growth

Over the past decades, China and ASEAN have undergone a profound economic and social transformation. This has largely been attributed to their openness to international trade and foreign investment. Although this region has been affected by the recent global financial crisis, it is assumed that if energy supplies are abundant enough, this region, especially some of its emerging economies, will continue to enjoy economic growth and progress over the long term. China and Vietnam are expected to continue to grow faster than other economies in the region, followed by Indonesia, the Philippines, and Malaysia. China now, as the world's second largest economy, will gradually shift from an investment/export-oriented economy to an import/consumption-oriented economy over the medium to long term and see a gradual productive population fall amid the ageing of its population. It is estimated that the average annual growth of China and ASEAN-10 countries will be 7 and 6 per cent respectively for 2015.[1] These years find that China's economy has cooled down as policymakers in Beijing have sought to put growth on a stable and rational footing. But China's slowdown is a "healthy" and "intentional" adjustment as the central government works to rebalance the economy.

ASEAN is expected to grow as a giant consumption market with its population of 600 million. At the same time, rising labour costs and emerging potential risks in China will make ASEAN's cheaper and abundant labour more valuable as a production base. Less-developed ASEAN countries are about to launch economic development, invigorating the entire ASEAN economy. ASEAN is assumed to grow at an annual rate of 4.6 per cent through 2035.

Growth in energy demand is closely correlated with growth in per-capita income. Rising income will continue to lead to increased demand for goods that require energy to use and to produce, such as cars, refrigerators, and air conditioners. Based on assumptions of the International Energy Agency (IEA), China's GDP per capita is to increase 5.5 per cent per year, from US$6,000 in 2011 to US$21,000 in 2035, while ASEAN's is to increase at 3.7 per cent per year, from around US$3,700 to US$8,700 during the same period (see Figure 2.1).

FIGURE 2.1
GDP Per Capita in Selected Countries

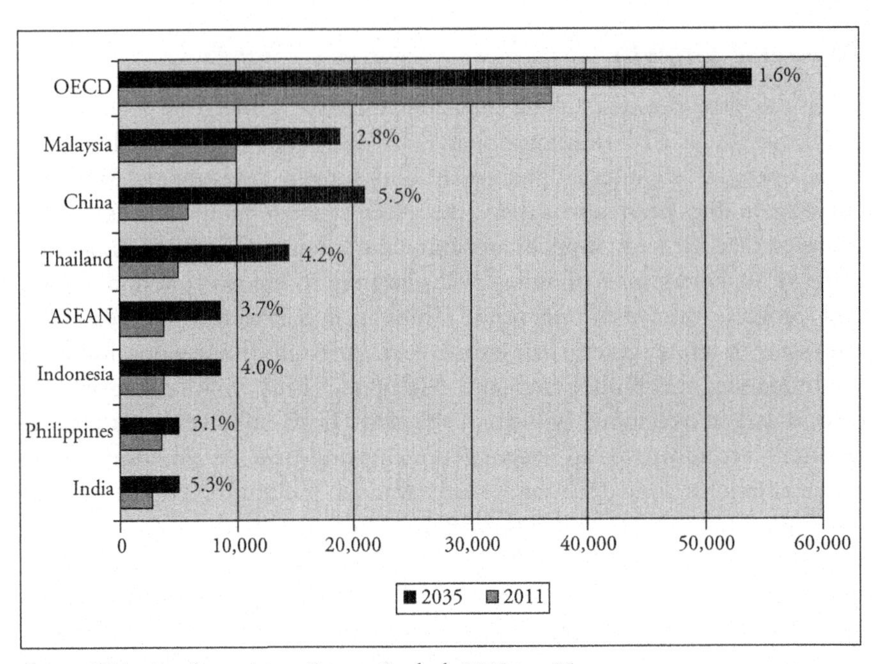

Source: IEA, *Southeast Asian Energy Outlook 2013*, p. 35.

Population Growth

Population and GDP growth are assumed to be the most significant drivers of energy demand in China and ASEAN. In 2010, China's population was 1.34 billion and is projected to reach 1.38 billion in 2020. Between 2000 and 2010, Southeast Asia's population (excluding Singapore and Brunei) grew at an average rate of 1.4 per cent per annum (reaching 586 million), a level that now exceeds that of the European Union.[2] Essentially all the increase in the population of these countries will occur in urban areas, while the population of rural areas is expected to decline over the period. For example, between 1990 and 2010, China's population living in urban areas grew at an annual rate of 2.5 per cent, the proportion of its total population living in urban areas has more than doubled, from 27 per cent in 1990 to 45 per cent in 2010,[3] with a target of 51.5 per cent by 2015 set out in the Twelfth Five-Year Plan. In ASEAN, it grew at 3.1 per cent between 1990 and 2010.

This urban population expansion will influence the growth in energy consumption in terms of aggregate level and energy structures. While the concentration of activities in urban areas can enable improved energy efficiency through economies of scale, the urban population in developing countries typically uses more energy, particularly in the residential and transport sectors, than their rural counterparts, as their higher incomes and better access to energy services typically outweigh energy efficiency gains that come from higher density of settlement.[4] For example, on the basis of the living standards in 2002, if the Chinese urban population increases 100 *yuan* per annum per capita in their expenditure on accommodation, the indirect energy consumption generated would increase by 20 million tons of standard coal, 10.15 per cent of the total indirect energy consumption generated in the daily life of the urban population in 2002.[5] This urban population expansion will also influence the energy structures, switching away from direct use of coal and biomass toward clean and refined fuels such as electricity and gasoline in the residential, commercial, and transport sectors.

Transportation Expansion

The expansion of the transportation sector, especially the growth in the number of motor vehicles, is another driving factor for increasing energy demand. According to the figures provided by Asian Development Bank (ADB), the number of vehicles in China had risen from 12.8 million in 1998 to 62.1 million in 2009.[6] During the same period, the number of vehicles in some ASEAN countries had also increased dramatically. The dramatic rise in the number of cars in China has been a controversial phenomena. It is significant that the Chinese authorities still think along the lines of improving the fuel-use efficiency of passenger cars and motorcycles, instead of limiting the increase in their numbers. This is because the automobile industry is considered a pillar industry in the Chinese economy; and the Chinese authorities now attempt to maintain economic growth through the stimulation of domestic consumption. The acquisition by Chinese families of passenger cars and motorcycles is still perceived as an important growth point in direct and indirect domestic consumption.

Energy Demand

Asia is at the centre of dramatic changes underway in world markets, as the region's energy demand booms in order to fuel dynamic economic growth and rising standards of living. The *2012 World Energy Outlook* by

IEA predicts that global energy demand will increase by one-third from 2010 to 2035, with Asia accounting for nearly two-thirds of that growth.[7] China and India alone will account for half of global demand growth. China, only recently established as the world's largest energy consumer, accounts for nearly 40 per cent of world energy demand growth from 2011 to 2035, and is projected to consume 4,000 Mtoe in 2035 (see Table 2.1).

Demand for energy in Southeast Asia has also grown significantly, although at a slower pace relative to China. Since 1990, ASEAN's energy demand has expanded two and a half times, accounting for 20 per cent of global energy demand growth in the past five years. By 2011, ASEAN's energy demand had reached 549 Mtoe, or around three quarters of that of India. As the economy grows in this region, ASEAN's energy demand will continue to rise sharply over the next two decades. As predicted by IEA, ASEAN's total primary energy demand will increase by 83 per cent, from 549 Mtoe in 2011 to 1,004 Mtoe in 2035 (see Table 2.1). As for individual member countries, Indonesia, Thailand, Malaysia, and the Philippines are the four largest energy consuming countries in ASEAN, while Vietnam was the country with the fastest growth rate in

TABLE 2.1

Primary Energy Demand in China and ASEAN

(Mtoe)

	1990	2000	2011	2020	2030	2035	2011–35
Asia	1,578 (18)	2,220 (22.0)	4,324 (33.1)	5,548 (36.9)	6,584 (39.6)	7,045 (36.9)	2.1%
China	879 (10)	1,175 (11.7)	2,743 (21)	3,519 (23.4)	3,945 (23.7)	4,060 (23.4)	1.6%
India	317 (3.6)	457 (4.5)	750 (5.3)	971 (6.5)	1,336 (8.0)	1,539 (6.5)	3.0%
ASEAN	223 (2.6)	373 (3.7)	549 (4.2)	718 (4.8)	897 (5.4)	1,004 (5.8)	2.5%
World	8,769 (100)	10,071 (100)	10,071 (100)	13,070 (100)	16,623 (100)	17,387 (100)	1.2%

Source: IEA, *World Energy Outlook 2013*, p. 69.

energy consumption over the last few years. Collectively, they accounted for 78 per cent of ASEAN's total energy consumption.

By energy type, fossil fuels will dominate energy demand for decades. Oil will remain the dominant fuel, with a share in 2030 of 20 per cent in China and 27 per cent in Southeast Asia. More than half of the incremental oil growth will be derived from the transport sector due to continued motorization along with economic development. The other sectors, including the agricultural and residential which will continue to use petroleum products such as kerosene and liquefied gas, are likely to contribute to the incremental growth as well. Coal will remain another important fuel for China and ASEAN, although its share in China's energy mix has been decreasing. The strong increase in coal demand is driven by its relative abundance in Southeast Asia and low coal prices, which lead to coal being favoured over (or substituted for) oil and natural gas, particularly in power generation in Southeast Asia.[8]

ENERGY SECURITY CONCERNS OF CHINA AND ASEAN

Defining Energy Security

Energy security, broadly defined, means adequate, affordable and reliable supplies of energy resources. It matters because energy is essential to economic growth and human development, especially for developing countries. Yet, no energy system can be entirely secure in the short term, because disruptions or shortages can arise unexpectedly, whether through sabotage, political intervention, strikes, technical failure, accidents, or natural disasters. Moreover, in the longer term, under-investment in crude oil and gas production, refining or transportation capacity, or other market failures can lead to shortages and consequently unacceptably high prices. So for many countries, energy security, in practice, is best seen as a problem of risk management, that is reducing to an acceptable level the risks and consequences of disruptions and adverse long-term market trends.

Concerns about energy security have evolved over time with changes in the global energy system and perception about the risks and potential costs of supply disruptions. In the 1970s and 1980s, the focus was on oil and the dangers associated with over-dependence on oil imports. Today, worries about energy security extend to natural gas, which is increasingly traded internationally, and the reliability of international energy security system, such as IEA. Together with oil and gas, which are carbon-producing fuels, is the worry about climate change and environmental damage. Post-Copenhagen climate change and environmental protection policies are likely to require real change in emissions in the medium and long term, and require more cooperation from developing countries.

Chinese scholars define energy as an important physical base of economic development, social progress, and the construction of modern civilization. For many Chinese scholars, energy is defined as a strategic material and major element of a country's security, which links national and foreign security policies.[9] Other Chinese scholars like Xu Xiaojie, a researcher of the Chinese Academy of Social Sciences (CASS), claims that the relation between energy and national security is subtle but significant. Xu emphasizes that "energy security is national security". He identified crude oil as a strategic commodity that is "indispensable for core functions of modern economic systems (and national defence)" and cannot be substituted in the short of medium term. Any state therefore has to identify the main risks to its oil supplies and assess the likelihood of the occurrence of events that can interfere with the uninterrupted supply of oil at acceptable prices. The extent to which such risks can be reduced depends on their nature and the means required in order to effectively deal with them. The assessment of such risks depends on a range of variables, such as the oil import dependence ratio, the oil intensity of the economy as well as the total level of imports, the security and diversity of transit routes, the diversity of sources of supply, the risks to supplier countries and the risks of natural disasters.[10]

Energy security, for consumers, has therefore generally meant the uninterrupted physical availability at a price which is affordable, while respecting environment concerns. In other words, the goal is to expand supplies fairly smoothly to match expected trends in demand, at prices to which importers have become accustomed, with environmental and climate constraints. Therefore, for developing Asian consumers like China, India, and some ASEAN countries, there are three aspects of energy security which are worth discussion: security from physical disruptions of supply, security from economic damage, and security from environmental damage. Based on the above definition and analysis, energy security concern of China and ASEAN countries can be specifically analysed on the following aspects.

Increasing Reliance on Oil and Gas Imports

Oil

With limited reserves and relatively flat domestic production as discussed in this chapter, China currently relies on the international markets for over half of the oil it consumes. China's vulnerability to an oil crisis is attributed to its rapid economic growth, high dependence on the Middle East for oil, and its backward oil reserve system. China is already second largest oil importer only after the United States (see Figure 2.2), and it will soon become the largest one. However, unlike the United States which has more diversified energy resources and imports most of its oil from Canada and the Gulf of Mexico, over 70 per cent of China's imported oil comes from the Middle East and Africa.

Moreover, the U.S. reliance on imported oil has been declining from about 70 per cent in the 1990s to below 40 per cent in 2012, and this figure will further fall to below 20 per cent beyond 2020 and toward zero net imports in 2030 as domestic energy production expands, biofuel and shale gas output increases, and energy efficiency improves.[11] Whereas, China's oil import dependence is reaching 55 per cent, well above the critical

FIGURE 2.2
Top Ten Net Oil Importers, 2013
(million barrels per day)

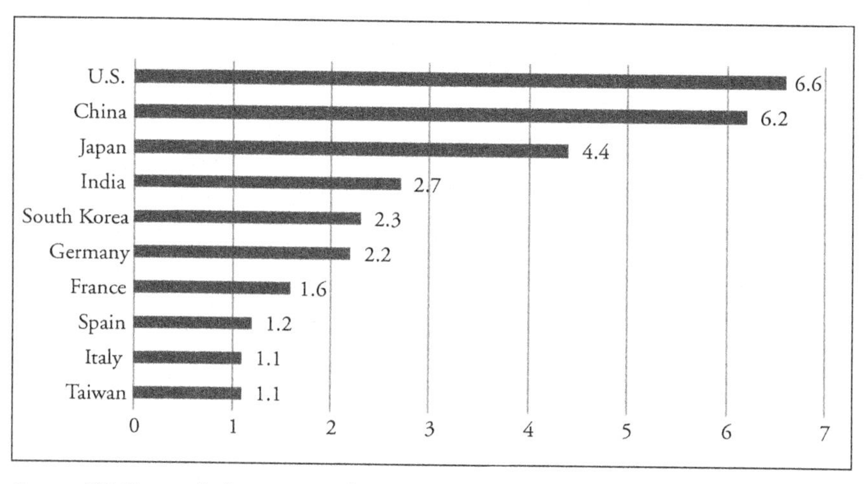

Source: US Energy Information Administration.

level based on international standards,[12] and is likely to increase to 80 per cent as predicted by International Energy Agency (see Figure 2.3). With China's increasing dependence on overseas energy resources, the economic risk posed by a shortage of energy resources is still the greatest risk for China's economic security.[13]

China's oil reserve system is still lagging behind that of the Western countries. A strategic oil reserve system was introduced in 1973 with the outbreak of Middle East war. The IEA required its member states to have at least 60 days reserves of oil, which was later increased to 90 days after the second oil crisis in the 1980s. Currently, the strategic oil reserves of the United States, Japan, Germany, and France are all above 90 days. The United States tops the list with oil reserves of 727 million barrels, and oil reserve days of more than 150 days.

In 2011, China imported 250 million tons of crude oil, an increase of 6 per cent over the previous year. China's current oil reserves are sufficient for only 30 days of use although the country has been making progress in this regard. China started building its oil reserve system in 2000, and by January 2010, four oil reserve bases (Zhenhai, Zhoushan,

FIGURE 2.3
Oil Import Dependence, China and ASEAN
(% of total oil consumption)

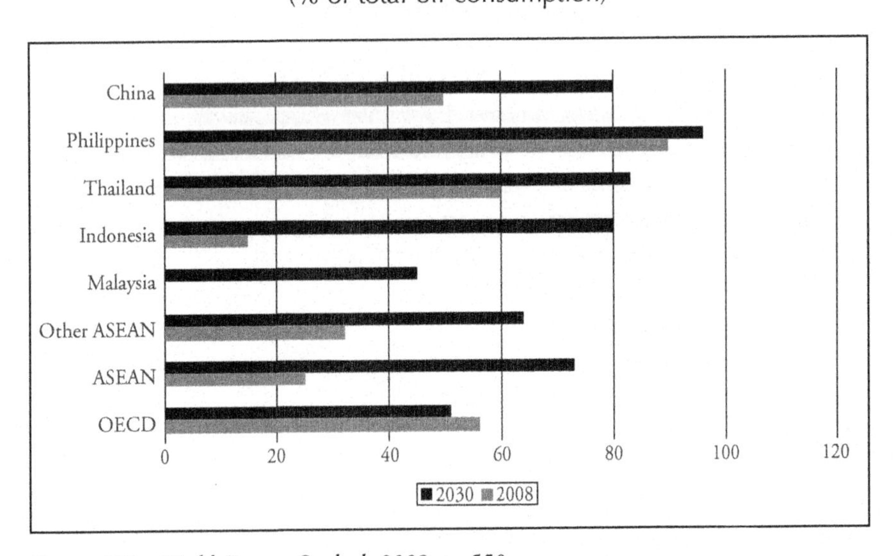

Source: IEA, *World Energy Outlook 2009*, p. 550.

Huangdao, and Dalian) have been put into use, with total oil reserves reaching about 100 million barrels. The second phase of the construction of an oil reserve base, which was completed in 2012, has increased China's oil reserves to 274 million barrels. China plans to complete the building of the whole strategic oil reserve system that has a storage capacity of 500 million barrels of oil reserves, or an equivalent to 90 days of China's oil imports, by 2020.[14]

The Middle East remains the largest source of China's crude oil imports, although African countries began contributing more to China's imports in recent years. China is heavily reliant on the unstable Middle East, being affected by geopolitics to a large extent. In 2011, China imported 179 million tons of oil, about 45 per cent of which comes from the Middle East, 20 per cent from Africa, 8 per cent from Russia, and 30 per cent from the United States and other regions (see Figure 2.4).

FIGURE 2.4
China's Crude Oil Imports by Source

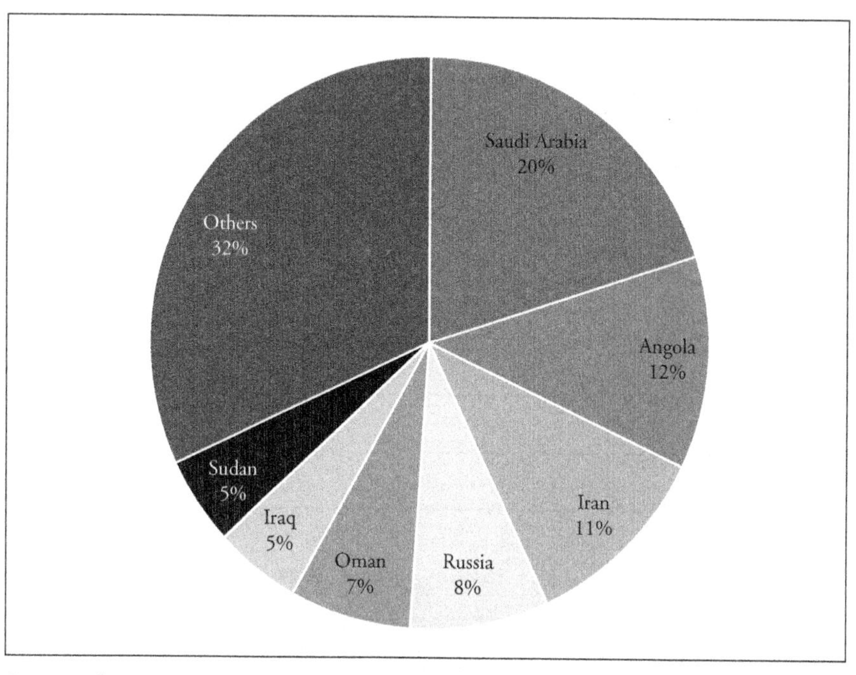

Source: China's Customs Statistical Yearbook 2011.

The energy situation of ASEAN is better than China, but energy security concerns have also been increasing in many countries. Southeast Asia has large reserves of oil and natural gas and has long played an important role as an exporter of both resources. By the early twentieth century, crude oil was being exported from what was then British Burma and Sarawak. However, this historical pattern is changing as the region's energy demand continues to grow, while oil output has been falling steadily, since peaking at around 3 mb/d (million barrels per day) in the 1990s. As a group, ASEAN has turned from a net oil exporter to a net oil importer since 1993, nearly in the same year as China. With the exception of Brunei and Malaysia, whose demand is expected to overtake production by the end of this decade, all ASEAN countries are currently net oil importers. ASEAN's oil production was 2.5 mb/d in 2012, and is projected to drop to 1.7 mb/d in 2035. With oil demand expected to continue to grow across Southeast Asia, declining output means that imports will continue to increase. ASEAN's net imports of oil are projected to increase by two-and-a-half times from 2012 to 2035, from 2 mb/d to over 5 mb/d (see Figure 2.5).

FIGURE 2.5
ASEAN's Oil Balance

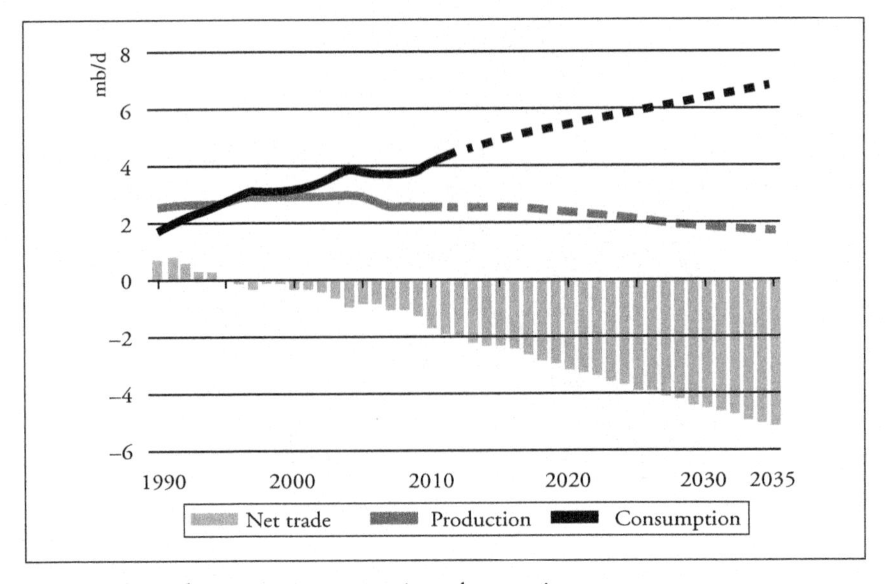

Note: Positive values are exports; negative values are imports.
Source: IEA, *Southeast Asian Energy Outlook 2013*, p. 83.

As a result, ASEAN has become the world's fourth largest oil importer, behind China, India, and the European Union, and its overall energy needs will rise to one billion tons of oil equivalent in 2035 from 549 million tons of oil equivalent in 2012.

Like China, ASEAN is also increasingly reliant on oil supply from the Middle East that is transited through the narrow Malacca Strait. For example, the percentage of ASEAN's crude oil imports from Middle Eastin its total oil imports had increased to over 70 per cent in 2009 (see Figure 2.6).

As for individual countries, Indonesia suspended its OPEC (Organization of Petroleum Exporting Countries) membership in 2008 as its outputs of oil and gas sharply dropped due to ageing oil and gas fields. In particular, the decline in crude oil production, combined with increasing domestic demand, turned Indonesia into a net oil importer in 2004. With projected growth in domestic oil demand, Indonesia's net import will more than double from 620 kb/d (thousand barrels per day) in 2012 to around 1.4 mb/d in 2035, and its oil import dependence will increase to over 80 per cent.

FIGURE 2.6
ASEAN's Crude Oil Imports by Source

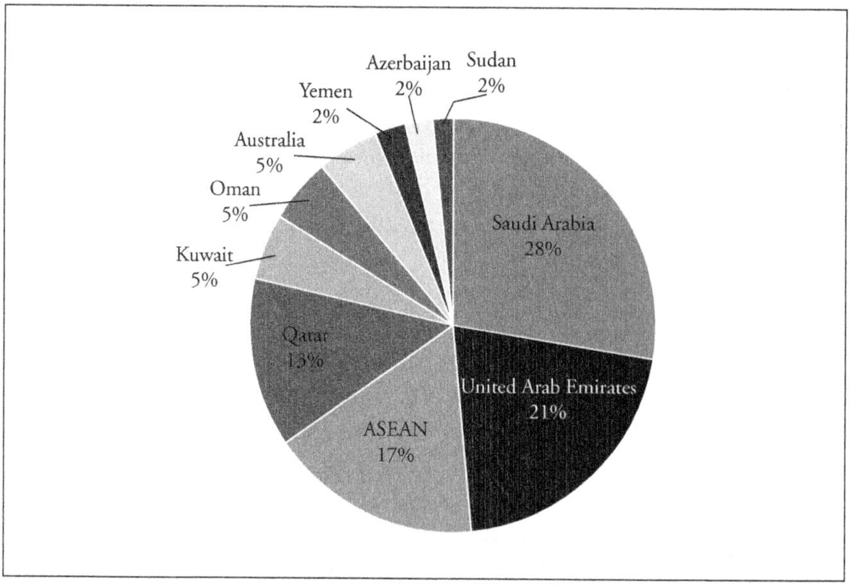

Source: *ASEAN Statistical Yearbook 2010*, p. 235.

Although Malaysia is the second largest oil producer in ASEAN, with an output of 670 kb/d in 2012, its oil production has seen a steady decline from a peak of 830 kb/d in 2003 as major producing fields have matured. Its output is expected to decline to around 419 kb/d by 2035 and its imports will reach 300 kb/d in 2030. By that time, Malaysia's oil import dependence will be around 45 per cent.[15]

In 2012, Thailand produced 440 kb/d of oil, while it consumed 1,212 kb/d, leaving total net imports of 772 kb/d, and making the country the second largest net oil importer in Southeast Asia only after Singapore, and by 2030 its oil import dependence will also rise to over 80 per cent. Around 95 per cent of Thailand's crude oil imports came from countries in the Middle East, including the United Arab Emirates, Saudi Arabia, Oman, Qatar, and Yemen.

Vietnam has rich energy resources, such as coal, oil, and natural gas, and is a net energy-exporting country right now. However, due to high economic growth and rapid urbanization, the country's energy demand will increase 2.5 times in 2015 and 5 times in 2025 compared with present consumption levels, even if the promotion of energy conservation is considered.[16] Moreover, part of its oil and gas are produced in the disputed areas of the South China Sea, bearing potential political risks. Vietnam is thus expected to become a net oil importer in around 2015.

The Philippines is the fifth largest energy consumer in Southeast Asia after Indonesia, Thailand, Malaysia, and Vietnam. Oil accounted for 34 per cent of the Philippine total primary energy consumption in 2009, and the country's oil production in 2008 was 23 kb/d. As Philippine total oil demand was 234 kb/d in 2008, imports of 211 kb/d were required, accounting for 90 per cent of its total oil consumption. Philippine total oil production is projected to decline gradually to around 10 kb/d in 2030, and its imports will rise to 235 kb/d in 2015 and 400 kb/d in 2030.[17]

Gas

China plans to increase national gas consumption in the next two decades as it strives to reduce dependence on coal and cut greenhouse gas emissions. But in spite of ambitious plans to exploit domestic unconventional gas, China will increasingly rely on imports to meet projected demand.

In 2013, China consumed 167.6 billion cubic metres (bcm) of natural gas, accounting 5.9 per cent of the national energy mix, of which about 31.6 per cent was imported.[18] In 2011, 29 per cent of China's liquefied natural gas (LNG) import came from Australia, 17 per cent from Indonesia, and 13 per cent from Malaysia (see Figure 2.7). It plans to increase the share of gas to 12 per cent in its total energy mix by 2030. The China National Petroleum Corporation (CNPC) and the China Petrochemical Corporation (SINOPEC) project that Chinese gas demand will rise at an average growth rate of around 8 per cent per year between 2012 and 2030.[19] It is predicted that the total demand will be 550 bcm/y by 2030, accounting for around one third of total global gas demand growth during this period.[20] Shale gas potentially gives China a previously unforeseen domestic option for improving its supply security. Given that China is home to the largest volume of technically recoverable shale gas resources in the world at 1,115 trillion cubic feet,[21] it hopes to develop its unconventional resources to reduce its import dependence. However,

FIGURE 2.7
China's LNG Import Sources, 2011

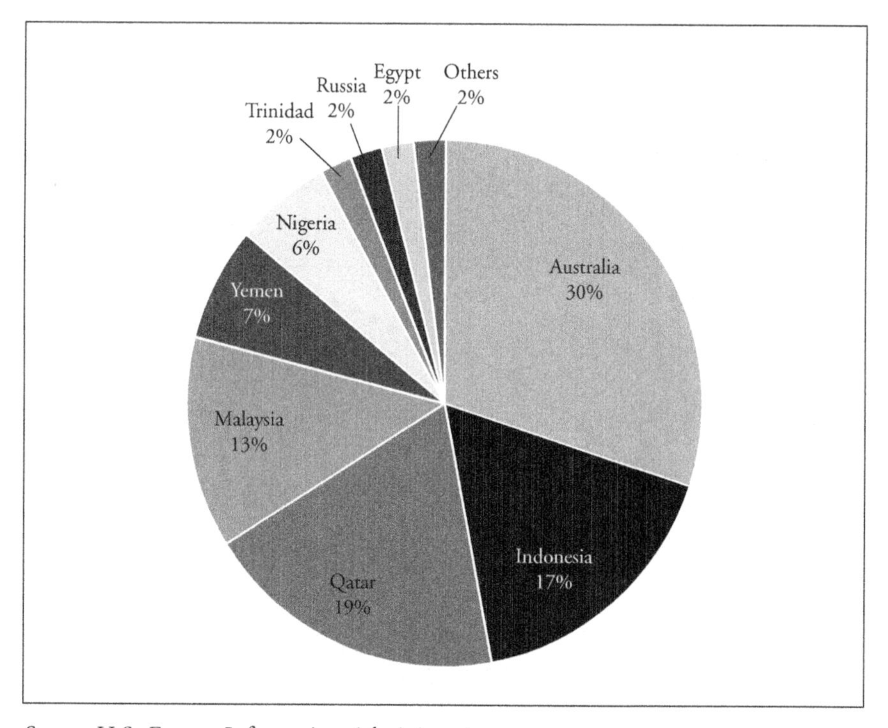

Source: U.S. Energy Information Administration.

China is at an early stage in unlocking its shale resource potential and faces challenges stemming from a lack of adequate technology and technical expertise, necessary infrastructure, water (in some locations), and market conditions.

China will rely on existing and under-construction pipelines from Central Asia and Myanmar, but these, together with current levels of domestic production, can only supply at most 40 per cent of demand. The remainder will have to be met through a combination of LNG, Russian pipeline gas, and new domestic reserves of unconventional gas.

For ASEAN countries, while the region remains an important supplier of LNG, gas is also increasingly sought to support power generation and industry in domestic markets. Net exports from the region are expected to increase in the medium term, approaching 70 bcm in 2020, and then incline to 14 bcm in 2035 and domestic demand increases (see Figure 2.8).

FIGURE 2.8
ASEAN's Gas Balance

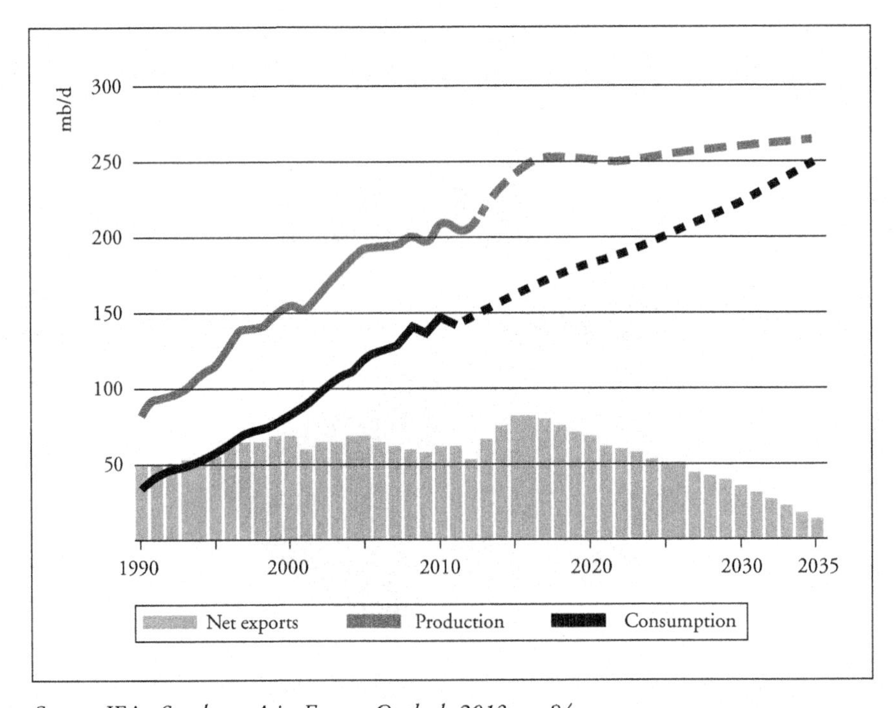

Source: IEA, *Southeast Asia Energy Outlook 2013*, p. 84.

Sea Lanes Security

Indian Ocean

The Indian Ocean is nearly 10,000 km wide from west to east at the southern tips of Africa and Australia, respectively, while extending nearly 13,500 km north to south, from the Persian Gulf to Antarctica. It is encompassed by a land rim on three sides — west, north and east, with maritime access to the region possible through seven established gateways or choke points. To the west, the Strait of Hormuz connects the Persian Gulf to the Indian Ocean and the busiest shipping lane passes through it. To the east, the Malacca Strait is the primary route which connects the South China Sea to the Indian Ocean and through which more than 50,000 vessels transit annually. Overall, the Indian Ocean accounts for the transportation of the highest tonnage of goods in the world, with almost 100,000 ships transiting its expanse every year, carrying two-thirds of the world's oil shipments, one-third of the bulk cargo traffic, and half the world's container shipments.[22] The Indian Ocean came into focus with the discovery of oil in the Persian Gulf. The region became strategically important with the first oil shock in the mid-1970s.

The importance of the Indian Ocean as an energy transportation corridor will further increase as global energy needs are likely to increase by 45 per cent between 2006 and 2030, and over half of this growth is to be from China, India, and emerging ASEAN countries. More than 85 per cent of the oil and oil products bound for China cross the Indian Ocean and pass through the Straits of Malacca. The phenomenal economic growth in the developing countries of China, India, South Korea, and the ASEAN nations has brought about a transformation in Asia. According to the United States National Intelligence Council analysis in 2008, the United States will find itself as one of a number of important actors on the world stage by 2025, but still the most powerful one. Militarily and strategically, the United States will continue to have an overall edge over all other states. Although advances by others in science and technology, irregular warfare, proliferation of long-range precision weapons, and the growing use of cyber warfare will constrict the United States freedom of action, the U.S. Navy will remain the most powerful in the Indian Ocean through the coming decades. It will also continue to be robustly postured in the Indian Ocean/Arabian Sea and the Western Pacific, as clearly stated in the U.S. Joint Maritime Strategy document brought out in 2007.[23]

South China Sea

Southeast Asia is essential as a transit area for China's energy resource supply, since four sea lanes in the South China Sea area are used to connect the country with the Middle East and Africa, and even with Latin America. The most important is perhaps the Straits of Malacca, which is however restricted to tankers under 100,000 tons or so;[24] the second is the Sunda Strait[25] for shipments from the Middle East and Africa; then the Philippine Sea for shipments from Latin America and the South Pacific; and then through the Lombok and Makassar Straits, or Lombok and Maluku Straits to connect the Philippine Sea to China.[26] Among these shipping routes, the most vital sea lane is the Straits of Malacca. In this strategically significant region, China's energy security "suffers from gaping vulnerabilities due to its excessive dependence on this one particular bottlenecked sea-lane".

Most of China's imports and exports go by sea, particularly trade with the European Union, the Middle East, and Africa, which has to travel through the Straits of Malacca before entering the South China Sea. Even China's crude imports shipped from Venezuela and South American countries have to pass either through the Philippine archipelago or between the Philippines and Taiwan (Luzon Strait). The rest of China's crude imports have to follow the route of the Indian Ocean through the Straits. By 2011, the shares of China's crude oil imports from the Middle East and Africa constituted 45 per cent and 20 per cent, respectively, of total oil imports, and it is predicted by the IEA that the proportion of China's oil imports coming from the Middle East will rise to at least 70 per cent by 2015.[27]

Given that oil is intimately related to China's economic development and sociopolitical stability, and that around 80 per cent of China's oil imports come from the Middle East and Africa, all but 10 per cent of the oil on foreign-owned tankers headed to China passing through U.S.-patrolled laneways of the Indian Ocean, into the Straits of Malacca and through the South China Sea, the fear of interdiction of China-bound oil tankers by the U.S. Navy is acute and real, with many Chinese strategists recalling Imperial Japan's vulnerability to maritime strangulation of its oil imports by the United States and allied navies in the Pacific Ocean leading up to World War II.[28]

Although this scenario is imaginable and would only happen in the event of a major war between the United States and China, or Japan and

China, Chinese security analysts remain uneasy at the prospect of U.S. dominance in the Indian Ocean and South China Sea. They are concerned that the United States and Japan are using the excuse of territory disputes as pretext to expand their naval presence in and around the South China Sea. Chinese strategists continue to argue that U.S. influence over key maritime trade routes, backed by security partnerships with many of China's neighbours in East Asia, represents a key threat to the security of China's maritime trade, including energy imports.[29]

Indeed, if we look at seascapes to the east of China, we can see what strategists call the "First Island Chain". That virtual arc goes from Japan and the Ryukyu islands and the Korean peninsula in the north, moving southwards via Taiwan, the Philippines, and Indonesia toward Australia. Assuming any serious confrontation along this arc, the U.S. Navy will be able to move its aircraft carriers around and seriously compromise China's access to its oil transported via the Malacca Straits. So from the point of view of Chinese naval strategists, the "First Island Chain" is deployed as a sort of "Reverse Great Wall" which is like an invisible sea wall from Japan to Australia that can, in theory, block China's access to the Pacific.[30]

Moreover, recent years have witnessed that the United States is very publicly repositioning the majority of its fleet to the Asia Pacific, and tensions are mounting between China and U.S. allies with Japanese Prime Minister's Shinzo Abe's proposed "democratic security diamond", a multilateral strategic alliance ultimately aimed at encircling China. To ensure peace in the high seas between the Indian Ocean and the Western Pacific and to curb China in the region, Prime Minister Abe encourages greater cooperation among Australia, India, Japan, and the United States through its bases in Hawaii. It also critically involves India, which has a larger and more capable navy than China and which increasingly sees itself as a guarantor of sea lanes not only in the Indian Ocean but also beyond into Southeast Asia. And in Southeast Asia itself, the ASEAN regional institution provides a strategic capacity for the smaller states of Southeast Asia to act as a unified actor so as to increase their bargaining power towards China.[31]

In this sense, the rising tensions in the South China Sea and the more active U.S. role in those disputes have complicated China's concerns on energy security, being mixed with Chinese debates over security issues and sovereignty issues in disputed maritime claims. Adding to the mix have been China's tougher stance in defence of its interest in disputed maritime regions. The overall impact of China's assertive approach regarding maritime disputes on the energy security debates in China remains to be determined, though it would appear for now to reinforce nationalist perspectives seeking to guard Chinese interests in the face of perceived adverse foreign initiatives and encroachments.

The Straits of Malacca is as well critical to the energy security of ASEAN countries, given an increased demand for imported oil and gas, most of which is expected to come from the Middle East. In addition to this, the Straits of Malacca and the Sunda Strait also carry a significant amount of container traffic: large ports sit astride both of these sea lanes. The ports that lie along the Malacca and Singapore Straits include not only Singapore but also Malaysia's primary port, Port Klang, and Tanjung Pelepas. In Indonesia, Tanjung Priok sits close to the Sunda Strait. Piracy, terrorism, or any other disruptions are major threats to the security of shipping in the sea lanes of Southeast Asia. With pirate activities growing in the straits across 2003–4, the possibility of transnational terrorist groups disrupting the maritime traffic became tangible. Since then Indonesia, Malaysia, Thailand, and Singapore have begun to coordinate air and naval patrols in the region. As a result of these and other initiatives, the number of pirate attacks in the area has declined.[32]

CLIMATE SECURITY

Climate Change and Its Impacts

Climate change, which is closely related to a country's economic growth and energy structures, is likely to be another significant challenge confronting China and ASEAN countries. A growing and increasing affluent population, together with further industrial growth, will result in more emissions in the future as well as greater share for the region in global emissions. Fossil-based energy and expansion of transportation have negative environmental consequences on both the production and consumption sides, contributing greatly to the greenhouse gas emissions and the changing climate. For example, the region's total CO_2 emissions have been growing at a faster rate than the global average because the region's GDP growth is higher. According to EIA, the share of China and ASEAN in global CO_2 emissions increased from 25 per cent in 2007 to 31 per cent in 2011 (see Table 2.2).

TABLE 2.2
CO_2 Emissions from the Consumption of Energy
(million metric tons; %)

	2007	2008	2009	2010	2011
China	6,326	6,685	7,573	7,997	8,715
Brunei	10	11	7	8	9
Cambodia	4	4	4	4	5
Indonesia	372	370	406	415	427
Laos	1	1	1	1	1
Malaysia	166	171	175	191	192
Myanmar	16	15	12	12	14
Philippines	74	74	69	78	81
Singapore	148	155	192	229	212
Thailand	247	355	268	273	270
Vietnam	90	103	105	121	113
Total	7,454 (25)	7,944 (26)	8,812 (29)	9,329 (30)	10,039 (31)
World	29,733 (100)	30,256 (100)	30,236 (100)	31,502 (100)	32,579 (100)

Source: EIA, *International Energy Statistics*, <http://www.eia.gov/cfapps/ipdbproject/IEDIndex3.cfm?tid=90&pid=44&aid=8http://www.eia.gov/cfapps/ipdbproject/IEDIndex3.cfm?tid=90&pid=44&aid=8 > (accessed 20 November 2013).

China and most ASEAN countries are resource-intensive economies. Their striking economic growth in the past years has been fuelled by massive and reckless development of fossil fuel-based energy systems, in ways that are clearly unsustainable. Table 2.3 lists the carbon intensities of China and ASEAN countries. These data point out that China and most ASEAN countries have relatively high carbon intensities when measured as CO_2 emissions per dollar of GDP. They emit more CO_2 than the average world level, much higher than the level of Japan, indicating more and more products of these countries are not able to meet ever-tougher emission standards for world trade.

TABLE 2.3
Carbon Intensity
(metric tons of CO_2 per 1,000 US dollar GDP)

	2007	2008	2009	2010	2011
China	2.2	2.1	2.2	2.1	2.1
Brunei	1.1	1.1	0.76	0.86	0.86
Cambodia	0.53	0.5	0.5	0.47	0.47
Indonesia	1.2	1.1	1.1	1.1	1.06
Laos	0.4	0.4	0.38	0.34	0.31
Malaysia	1.07	1.05	1.09	1.1	1.07
Myanmar	1.06	0.95	0.72	0.68	0.75
Philippines	0.64	0.62	0.67	0.6	0.59
Singapore	1	1.03	1.3	1.34	1.19
Thailand	1.27	1.28	1.38	1.3	1.25
Vietnam	1.45	1.57	1.51	1.63	1.43
Japan	0.26	0.26	0.25	0.26	0.26
World	0.6	0.6	0.61	0.62	0.62

Source: EIA, *International Energy Statistics*, <http://www.eia.gov/cfapps/ipdbproject/IEDIndex3.cfm?tid=91&pid=46&aid=31> (accessed 26 March 2015).

China and ASEAN countries have already started to encounter problems from climate change. Carbon intensive economy has caused significant environment impacts for China's urban and rural population. Air pollution costs 4.4 per cent of world GDP, with a high of more than 10 per cent of GDP in China, the top carbon emitter. According to the estimates presented by the National Development and Reform Commission (NDRC), nearly 750,000 people die each year of respiratory illnesses brought on or exacerbated by poor air quality; 700 million of the rural poor have inadequate access to safe water and energy infrastructure; and the quality of the water in 40 per cent of China's rivers is very low.[33]

Southeast Asia is particularly vulnerable to the impacts of climate change due to the concentration of people and economic activities in the coastal areas, its rich biological diversity, and its mainly resource-based economies. For ASEAN countries, climate change is especially manifested in temperature increase, extreme weather conditions, and rising sea levels.

Energy Structures and CO_2 Emissions

The increasing CO_2 emissions in China and ASEAN countries are closely linked to their energy structures on both supply and demand sides. The projected trends in rising energy demand mean that energy-related CO_2 emissions from countries in this region continue to increase. As the world's largest coal producing and consuming country, according to IEA, China's CO_2 emissions reached 6 billion tons in 2007 and are expected to rise to 9 billion tons by 2030; Southeast Asia's share of global CO_2 emissions will rise to 5 per cent in 2030, up from 3.5 per cent in 2007.[34]

Main Source of Emissions

Supply side

Energy consumption is the main source of CO_2 and climate change. China's and ASEAN countries' increasing appetite for energy has been driven by rapidly expanding power sector, particularly for China. Coal-fired thermal power plants are the main greenhouse gas emitter and the source for high carbon intensities. According to the *Asia/World Energy Outlook 2012*, of the present global electricity generation, coal accounts for the largest share at 41 per cent followed by 22 per cent for natural gas, 16 per cent for hydro and 13 per cent for nuclear.[35] As for China and ASEAN countries, in 2009, coal accounted for 79 per cent of energy sources for electricity production in China, 42 per cent for Indonesia, 31 per cent for Malaysia, and 27 per cent for the Philippines, while the shares of gas and renewable sources are relatively low (see Table 2.4). Worse, the share of coal is still increasing in Southeast Asia. For example, for ASEAN as a whole, coal accounts for 32 per cent for electricity generation in 2011, and this number will increase to 48 per cent in 2035.[36] Indonesia is set to lead the growth in Southeast Asia's coal demand, with its abundant coal resources and an already booming export sector. Coal demand in Indonesia has increased at 11 per cent per year for the last two decades.

TABLE 2.4
Electricity Production Sources of China and ASEAN Countries, 2009
(%)

	Coal	Natural Gas	Oil	Hydropower	Renewable Sources	Nuclear Power
China	78.8	1.4	0.4	16.7	0.8	1.9
Cambodia	0	0	95.6	3.9	0.5	0
Indonesia	41.8	22.1	22.8	7.3	6	0
Malaysia	30.9	60.7	2.0	6.3	0	0
Myanmar	0	19.6	8.9	71.5	0	0
Philippines	26.6	32.1	8.7	15.8	16.8	0
Singapore	0	81.0	18.8	0	0.1	0
Thailand	19.9	70.7	0.5	4.8	4.0	0
Vietnam	18.0	43.4	2.5	36.0	0	0

Source: The World Bank, *World Development Indicators 2012*, pp. 166–68.

China is the world's largest coal producing and consuming country. While expanding coal production is the cheapest and most secure option, it comes with rising external cost. Hence, China's import of coal has been rising while ASEAN has made a great shift from oil to natural gas as electricity generation fuel since the 1990s due to natural gas development in the Bay of Thailand and other locations. As gas production has hit a peak and gas demand emerged in other sectors, gas supply capacity for electricity generation has become short, and its share in electricity generation sources will decrease from 44 per cent in 2011 to 28 per cent in 2035.

Therefore, the ASEAN electricity generation mix sees a shift from gas to coal. According to IEA, through 2035, the total amount of coal consumption in ASEAN will increase to 270 Mtoe (see Figure 2.9). The power sector accounts for approximately 60 per cent of coal consumption today. In order to meet the rapid increase of electricity demand, ASEAN countries will need to replace oil and natural gas, which are more expensive fossil fuels in the international market, and use more coal which is cheaper and more abundant in Southeast Asia. Together with oil and gas, coal continues to serve as a mainstay electricity source for China, India, and many ASEAN countries. Hence, the overwhelming reliance on fossil fuels for electricity

FIGURE 2.9
ASEAN Energy Supply by Source
(Mtoe)

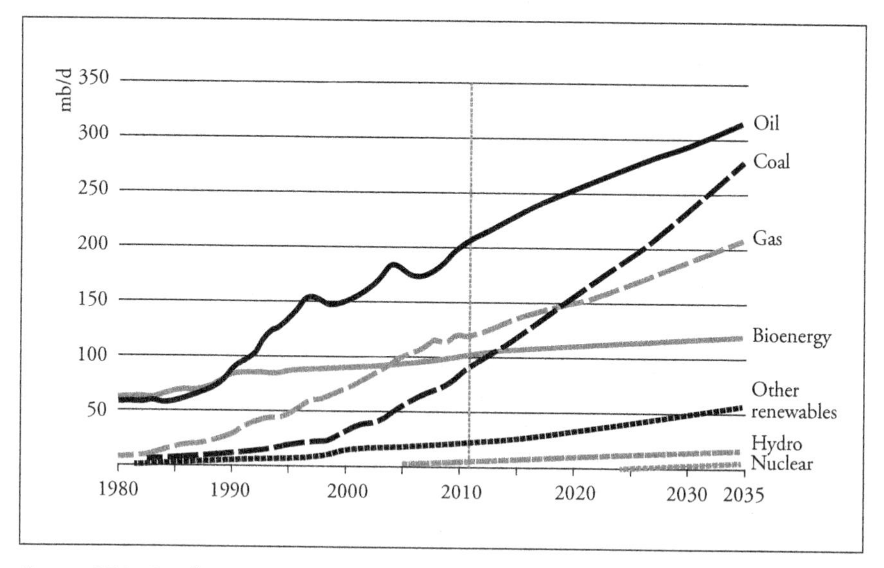

Source: IEA, *Southeast Asia Energy Outlook 2013*, p. 38.

generation will continue to incur the problem of significant greenhouse gas emission in Asia.

Demand side

Despite improvement in recent years, local pollution remains a major public health issue in many parts of China and Southeast Asia, particularly in major cities such as Beijing, Shanghai, Jakarta, Bangkok, Manila, Kuala Lumpur, and Ho Chi Minh City. This is linked primarily to rising vehicle use, rapid rates of industrialization and urbanization, the heavy reliance on coal, and siting of industry close to residential areas.

Table 2.5 shows that electricity generation and transport sectors are the main sources for CO_2 emissions, as China's electricity generation accounted 52 per cent in 2008. According to IEA, despite improvements in thermal efficiency, the power sector continues to be responsible for most of the increase in CO_2 emissions to 2030 and its share in total emissions remains broadly constant.[37] This is because of the fast growing demand for electricity and because the share of coal in the electricity mix is projected to remain high.

TABLE 2.5
CO_2 Emissions by Sectors
(% of total fuel combustion)

	Electricity and Heat Production		Manufacturing Industries and Constructions		Residential Buildings and Commercial and Public Services		Transport		Others	
	1990	**2008**	**1990**	**2008**	**1990**	**2008**	**1990**	**2008**	**1990**	**2008**
China	32.2	51.9	40.9	33.3	16.5	6.2	5.3	7.0	5.1	1.6
Cambodia	1.1	34.3	8.2	7.0	25.5	8.4	65.2	50.6	0	0
Indonesia	37.3	37.7	23.7	34.0	13.7	6.3	22.7	19.7	2.7	2.3
Malaysia	35.7	49.6	30.7	24.2	4.4	2.5	29.2	23.2	0	0.4
Myanmar	39.9	20.8	28.1	25.4	0.3	2.6	31.7	28.1	0	23.1
Philippines	35.5	44.0	21.0	18.0	6.7	5.7	34.9	31.3	2.0	1.1
Singapore	78.1	71.4	6.7	11.9	0.6	0.5	14.0	16.2	0	0
Thailand	36.5	41.1	18.8	29.2	3.1	2.8	34.5	22.3	7.1	4.6
Vietnam	27.7	29.3	32.6	35.1	11.2	9.5	24.2	24.5	4.0	1.6

Source: The World Bank, *World Development Indicators 2012*, pp. 178–81.

This situation is most prominent for China. According to World Development Indicators 2012, the power sector, which is mainly fuelled by coal, contributes 52 per cent of China's CO_2 emissions in 2009, and its share is projected to rise 54 per cent in 2030. If China continues on its "business as usual" path, the IEA predicts that China's CO_2 emissions will rise by 3.3 per cent annually over the next twenty-five years, resulting in China's emissions in 2030 being 11.7 billion tons, twice the predicted level of the United States' emissions (5.8 billion tons) that year.[38]

For ASEAN countries, power generation is also the major source of growth in CO_2 emissions, accounting for around 40 per cent of the total CO_2 emissions in Southeast Asia in 2008. Transportation is another major source as its emissions accounted for 27 per cent in 2008, and will increase to account for 50 per cent in 2030. This is primarily due to rising demand for individual mobility and freight. The industrial sector (comprising manufacturing such as iron and steel, chemicals, non-metallic minerals) accounted for 23 per cent.

Obviously, China and ASEAN countries and other emerging countries need to adjust their energy structures, find ways to slow down the rise in coal consumption, and mitigate its environmental impacts. Improving energy efficiency and developing renewable energy are part of the solution to boost supplies of cleaner energy. If tomorrow's cities are to be more sustainable, it is essential to implement binding energy-efficiency standards, distributed generation for electricity, and standards for housing. Actually, China and ASEAN countries currently use less renewables in their energy mix than the rest of the world, so there is a lot of room to grow the use of renewables.

RETHINKING OF CHINA'S AND ASEAN COUNTRIES' RESOURCE-INTENSIVE GROWTH MODELS

Resource-intensive Development

China and most ASEAN countries are resource-intensive economies. Their striking economic growth in the past years has been fuelled by massive and reckless development of fossil fuel-based energy systems, in ways that are clearly unsustainable. The costs, in environmental devastation, damage to health, and geopolitical instability, are high and growing, and increasingly unnecessary as new technologies reach a transformative tipping point.

Table 2.6 shows that China and most ASEAN countries have relatively high energy intensity (GDP per unit of energy use), as in China in 2009,

TABLE 2.6
GDP per unit of Energy Use
(constant 2005 PPP US$ per kilogramme of oil equivalent)

	2000	2001	2002	2003	2004	2005	2006	2007	2008	2009
China	3.1	3.3	3.4	3.3	3.3	3.2	3.3	3.5	3.6	3.7
Brunei	6.5	7.1	7.6	6.8	6.7	6.9	5.6	5.5	5.0	5.7
Cambodia	3.2	3.3	3.5	3.6	3.8	4.2	4.5	4.8	5.1	5.1
Indonesia	3.6	3.6	3.6	3.8	3.7	3.9	4.0	4.2	4.4	4.3
Laos	—	—	—	—	—	—	—	—	—	—
Malaysia	5.1	4.8	5.0	4.9	5.3	4.8	5.0	4.9	4.9	5.2
Myanmar	—	—	—	—	—	—	—	—	—	—
Philippines	5.2	5.5	5.7	5.9	6.4	6.7	7.1	7.7	7.7	7.9
Singapore	8.0	6.9	7.5	10.7	9.3	10.4	10.9	14.6	14.0	12.5
Thailand	4.8	4.7	4.6	4.6	4.5	4.6	4.8	4.8	4.7	4.8
Vietnam	3.4	3.4	3.4	3.5	3.3	3.5	3.7	3.7	3.8	3.7
Japan	7.1	7.2	7.2	7.4	7.3	7.5	7.6	7.8	8.1	8.0
World	4.9	5.0	5.1	5.1	5.1	5.2	5.3	5.4	5.5	5.5

Note: PPP = purchasing power parity
Source: World Bank, *World Development Indicators*, 2012.

per kilogramme of oil equivalent produced US$3.7 GDP which was lower than the average world level. China uses 21.5 per cent of global energy and generates 12.3 per cent of the world's GDP. It also indicates that over the past decades, only Singapore and Brunei performed better than the world average in efficient energy use, meaning that the remaining countries are less efficient than most in converting energy fuels into GDP. The problem of low energy efficiency appears most serious in China and Vietnam, which suggests that these countries urgently need to revise their development models to make economic growth less energy-intensive.

China's State-centric Economic Model

SOEs in China's Political Economy

China's resource-intensive economy has much to do with its state-centric economic model. During the early 1950s, the Chinese Communist Party (CCP) undertook a massive process of institutional transfer which laid down "a vast lattice work of Soviet derived political and economic institutions".[39] Mao Zedong and his supporters subsequently adjusted this Soviet development model to Chinese needs, setting up SOEs (State Owned Enterprises) as the "pillars" of China's national economy. This Maoist paradigm of development, which began with the convulsive "Great Leap Forward" in 1958 and culminated in the social, economic, and political anarchy of the Cultural Revolution (1966–76) left an ultimately tragic developmental legacy of famine and chaos. After 1978, Chinese paramount leader Deng Xiaoping embarked upon a course of economic reform, and the strategy of market socialism unleashed a process of unprecedented and rapid economic change in the 1980s and 1990s. This process has required the redefinition of the state's role in the economy in ways which, by the 1990s, brought the CCP leaders closer to the "state capitalist mode" to the extent that it disengages the state from direct economic control and increases the market accountability of productive enterprises.[40]

After over three decades of economic reforms and development, China appears to be a private-sector driven economy, but in fact, SOEs (state-owned enterprises) still play a dominant role in China's political economy. Currently (2014) China has approximately 150 central-level SOEs and 120,000 provincial-and local-level SOEs. This compares to four million private corporations and tens of millions of small, informal private businesses. SOEs have been the "pillars" of China's national economy since the founding of the new China, and the bias toward the state-controlled sectors is clear given the relationship between the state-controlled banks and industrial SOEs. For example, even though state-controlled enterprises produce 30–50 per cent of all output in the country, they receive over 75 per cent of the country's capital.[41]

China generally regards the SOEs' domination of every important emerging sector as a strategic priority in its economy. The Twelfth Five-Year Plan (2011–15) explicitly states that "national champions" are to take the lead in "strategic emerging industries" such as renewable energy, health care, biotechnology, high-end equipment manufacturing, energy-efficient vehicles, and information technology. It is made clear in the Plan that the government should "channel state capital and resources into industries essential to national security and economy though discretionary and rational capital injection". This mainly includes capital resources and preferential lending policies from state-owned banks.

This state-dominated set-up is reflected in the energy sectors. In the early years after oil self-sufficiency ended in 1993, Beijing reorganized its oil and gas assets and entities into two state-owned firms: the China National Petroleum Corporation (CNPC) and the China Petroleum and Chemical Corporation (Sinopec). CNPC is the dominant upstream player in the sector, and along with its listed entity — PetroChina — accounts for over 66 per cent of China's oil output. Sinopec accounts for at least half of the country's downstream activities such as refining and distribution. The state-owned China National Offshore Oil Corporation (CNOOC) is close to being a monopolistic player in offshore oil exploration and production, with other state-owned giants such as Sinochem Group becoming more prominent in offshore oil distribution.[42]

Hence, China's onshore and offshore commercial activities in oil and gas sector are dominated by these NOCs' (National Oil Companies). CNPC and Sinopec were in the top ten of the 2012 Forum Global 500 rankings.[43] The NOCs' domination of these sectors is enhanced not only through exclusive access to oil and gas assets, development, and distribution, but also through ongoing financing agreements with the state-owned banking sector, which offers the NOCs privileged access to plentiful credit. To ensure cheap and reliable energy for national economy remains the priority objectives of these NOCs.

Rapid Economic Growth and High Energy Demand

If we review China's economic development process in the past decades, we can find that high growth rate remains the main theme. For China, maintaining the economic growth rate stems from the fact that the modern CCP largely stakes its legitimacy on the capacity to continually deliver rapid economic growth, and can be realized under the state-centric economic model.[44] Former Premier Wen Jiabao explained explicitly during the National People's Congress in March 2011, "ensuring GDP growth of around 8 per cent each year and keeping inflation below 4–5 per cent is linked to social stability imperatives required for regime security".[45] In Beijing's view, such rapid growth is required to generate sufficient jobs to keep unemployment under control and maintain social stability. In such a politicized economy, Beijing cannot leave it to markets to determine supply, pricing, and distribution of energy resources.

The domestic structure of China's political economy means that it relies heavily on SOEs to achieve rapid growth rate and secure energy resource supplies. In the first decade of economic reform (1979–89), growth was largely driven by land reforms, which led to dramatic productivity increases in rural areas. In the mid-1990s, large-scale industrialization took place and became the driver of economic growth. Soon after China entered the World Trade Organization (WTO) in 2001, it began an expansion of infrastructure and fixed-asset investments that led to later property sector boom. To meet these macro-economic demands, China focused on the development of some heavy industries (such as steel, aluminum, and cement) that could drive high domestic economic growth. These industries largely ran on coal and depended on imported commodities such as iron ore and coking coal to operate. Examining Chinese oil consumption over the last decades makes this clear. From 2000–10, China's coal demand growth averaged 9 per cent per year, more than double the global growth rate of 4 per cent.

During the global financial crisis (2008–10), the leadership viewed large-scale stimulus efforts to sustain high rates of GDP growth as the key to political stability. The central government allocated 4 trillion RMB to infrastructure sectors and the majority part of these loans flowed to SOEs. Fixed-asset investment again became the main growth driver, driving over three-quarters of GDP growth. But fixed-asset investment is an immensely energy-intensive form of economic activity, especially in an economy that still uses energy extremely inefficiently compared to Western industrialized ones.

After Chinese new leadership under Xi Jinping and Li Keqiang took power in late 2012, China adjusted its development model, focusing more on the development of service sectors and urbanization. As China's economy moves from dependence on energy-intensive industrial manufacturing to a more service-oriented and urbanization-driven economy, the transportation sector becomes the most important source of growth in liquid fuels use. China's growth model is increasingly oil-dependent, and oil's main function is urban transportation. As analysed earlier in this chapter, China's population is about 51 per cent urban, with around 20 million people urbanizing each year, leaving scope for about 400 million to urbanize over the next twenty to thirty years (the figure is almost a third larger than the population of the United States). Thus until alternatives become affordable and mainstream, oil demand from China will be immense.

As a consequence of resource-intensive growth model, China has seven of the world's ten largest most polluted cities, and a new wave of urbanization over the coming ten years is projected to add 350 million more to the country's urban population. By 2025, there will be 221 Chinese cities with a million or more living in them. By comparison, Europe has thirty-five cities with one million or more in them. China's rapid urbanization contributes to continuing growth of energy demand, which helps explain why despite large scale investments in nuclear power, wind, and solar, 70 per cent of its energy still comes from coal, a percentage that has remained steady over the past two decades.

In short, the country's emphasis on high GDP growth rate, rapid industrialization, and recent urbanization over the last decades has had significant effects on Chinese energy consumption model, both in terms of scale and in the type of resources consumed. Industry in China continues to be responsible for the vast majority of energy consumption — primarily in the form of coal — and for most of the associated environmental degradation and emissions from consuming fossil fuels. Although the current Chinese government has been making great efforts to adjust its development model and economic structures, its current growth model still has an investment bias toward urbanization and massive infrastructures. This model is continuously largely powered by coal and fossil fuels, worsening environmental degradation and increasing undue pressure on limited resources. Such a growth model itself is unsustainable and will require a transition to a less energy-intensive and more consumption-driven economy.

ASEAN's Fossil-fuel Subsidy Model

Export-driven Development

Many Southeast Asian countries like Thailand, Malaysia, and Indonesia possess large rural and resource-rich hinterlands that facilitated their pre-war emergence as primary product economies. Decolonization and political instability that engulfed Southeast Asia from 1959 to 1975 adversely affected investment and economic development. These Southeast Asian states adopted economic nationalist and import-substituting strategies in order to acquire investment capital in the early 1960s and 1970s, much less successful than in Northeast Asia and Singapore. Indeed, it was only when they shifted to policies that favoured foreign multinational investment and export-led growth in the course of the 1980s that these later developing economies achieved rapid growth.[46]

Table 2.7 shows that for most ASEAN countries, exports of goods and services account for a high percentage of their GDP. More importantly, due to their comparative advantages, energy- and labour-intensive products account for the large part of their exports to developed countries, especially in the early stages of their economic development. However, energy-intensive industries take more fuel costs. For example, for food processing, textiles, lumber, paper processing, chemical manufacturing, and cement mixing in Southeast Asian countries, the costs of fuel account for more than 15 per cent of the final cost of the goods they produce.[47]

In order to help the exporting firms to be more competitive and keep exporting products to international markets, especially to the European Union and the United States, apart from paying low manufacturing wages and keeping low exchange rates, many ASEAN countries provided export subsidies. These subsidies were in the forms of subsidized loans for exporters, tariff rebates on export inputs, and subsidized energy. Many ASEAN countries provide subsidies that lower the price paid by energy consumers to below international market levels, thus reducing their product cost. The weakening of the currency would mean that the governments would have to spend more to buy oil on the international market. Consequently, fuel subsidies may impose a severe burden on the budget.

TABLE 2.7
Exports as a Percentage of GDP
(%)

	2000	2001	2002	2003	2004	2005	2006	2007	2008	2009	2010	2011
Brunei	67.4	69.5	67.1	69.3	68.8	70.2	71.7	67.9	78.3	72.8	81.4	81.3
Cambodia	49.9	52.7	55.4	56.5	63.6	64.1	68.6	65.3	65.5	59.9	54.1	54.1
Indonesia	41.0	39.0	32.7	30.5	32.2	34.1	31.0	29.4	29.8	24.2	24.6	26.3
Laos	—	—	—	—	—	—	—	—	—	—	—	—
Malaysia	119.8	110.4	108.3	106.9	115.4	112.9	112.2	106.9	99.5	91.4	93.7	91.6
Myanmar	0.5	0.5	0.3	0.2	0.2	0.2	0.2	0.2	0.1	0.1	0.1	0.1
Philippines	51.4	46.0	46.7	47.2	48.6	46.1	46.6	43.3	36.9	32.2	34.8	31.0
Singapore	192.3	187.8	188.8	207.4	219.3	229.7	233.4	217.7	233.2	198.6	207.2	209.0
Thailand	65.0	63.4	60.8	61.6	66.1	68.6	69.0	69.2	71.7	64.6	66.7	...
Vietnam	55.0	54.6	56.8	59.3	65.7	69.4	73.6	76.9	77.9	68.3	77.5	87.0
China	23.3	22.6	25.1	29.6	34.0	37.1	39.1	38.4	35.0	26.7	29.4	28.6
Japan	10.9	10.4	11.3	11.9	13.2	14.3	16.2	17.7	17.7	12.7	15.2	15.2

Source: ADB, *Key Indicators for Asia and the Pacific 2012*, p. 172.

Fossil-fuel Subsidies and Energy Security

Actually, Southeast Asia has a long history of providing subsidies, direct and indirect, on fuels for both consumers and producers. ASEAN countries, with the exception of the Philippines and Singapore, subsidize fuel and electricity prices. These subsidies are, in large part, directed at gasoline and diesel as well as socially sensitive products. Within ASEAN countries, government spending on subsidies is largest in Indonesia and Malaysia.[48] The Indonesian government fixed the price at a very low level — below 20 cents per litre — until 2005, with the budget bearing the cost of the difference between domestic and international oil prices. When the world oil price started to rise substantially in 2004, fuel subsidies became the main expenditure item in the budget, consuming some US$15 billion, or more than 20 per cent of total expenditures, in both 2004 and 2005.[49] According to IEA's analysis, "fossil-fuel subsidies are the main factors leading to resource-intensive development model, as they encourage wasteful energy consumption, burden government budgets and deter investment in energy infrastructure and efficient technologies."[50]

Most economists agree that fossil-fuel subsidies are a bad idea, as they promote a misallocation of resources in the economy, namely the over-consumption of fossil fuels.[51] They can be a burden on the public finances, hinder much-needed infrastructure development, and undermine a country's effort to diversify its energy intake. What's more, this waste increases global carbon emissions. According to the IEA estimates, globally, the cost of government subsidies for fossil fuels increased from US$311 billion in 2009 to US$544 billion in 2012. Much of this spending is wasted, as overconsumption of energy does not lead to higher levels of economic output but instead lower levels of efficiency. Other research suggests that most of this spending leads to big "dead weight losses", meaning lost economic efficiency as a result of government intervention.[52]

Indeed, spending on subsidies often becomes a serious burden on government resources for many ASEAN countries. Based on IEA estimates, fossil-fuel subsidies in Southeast Asia amounted to US$51 billion in 2012, despite progress that has been made in reducing the overall amount.[53] Most ASEAN countries still provide at levels above the world average. In 2011, the after-tax energy subsidies on petroleum products, electricity, natural gas, and coal amounted to a share of GDP that was 8.4 per cent in Brunei, 5.3 per cent in Indonesia, 7.2 per cent in Malaysia, and 4.7 per cent in Thailand, compared with a world average of only 2.7 per cent.[54] In terms of the ratio of energy subsidies to overall government budgets, the government of Malaysia had the highest ratio at 32.9 per cent, followed by 30 per cent and 20.9 per cent in Indonesia and Thailand respectively.[55] Malaysia, which spent an estimated US$8.5 billion on fuel subsidies in 2012, cut subsidies to petrol and diesel in September 2013 in a bid to reduce its budget deficit.

Subsidized energy prices in Southeast Asia are restricting investment in energy infrastructure and renewable energy sectors by depriving energy companies of the revenues needed for new investment. In Indonesia, for example, as fuel subsidies consume over 20 per cent of the state budget, it constrains the remaining fiscal allocation for infrastructure development. The 2012 IMF Country Report (No. 12/278) on Indonesia's economy also argued that while the cost of capital is declining, investment in infrastructure (including in energy-related infrastructure) remains relatively weak.[56] Consequently, Indonesia still suffers from basic infrastructure deficiencies in numerous public sectors, including clean water, sanitation, health, public transportation, communication, education, and electricity despite the booming economic growth in recent years.[57]

Doubtless, price controls are slowing down the expansion of generating capacity, and grid extensions and upgrades that are needed to raise the electrification and efficiency rates. For exporting countries, as long as they have more exports than imports, surplus budgets can easily offset the real-market input costs of fuel subsidies. But this model is possible to be unsustainable, as countries like Indonesia and the Philippines record larger trade deficits due to slowing exports and rising import cost because of the stagnant global economic growth.

Although there is now widespread recognition that subsidies are not sustainable and are having many unintended consequences, subsidy policies cannot be abolished overnight. Public opinion across the world and Southeast Asia is hostile to higher energy prices. Hence, there are real barriers to reform efforts. In particular, as subsidy policy is so politically sensitive in the region, the reform efforts are often dictated by political realities and electoral cycles. Despite recent reform efforts, notably in Indonesia, Malaysia, and Thailand, subsidies remain a significant factor distorting energy markets.

In summary, for the past decades, economic growth models in China and ASEAN countries were built on the assumption of cheap and stable resource supply. However, the situation has been changing. The World Economic Forum (2012) has called for a shift in business models to enable growth through resource efficiency and renewable energy. It argues that this is a core strategic imperative for any economy that intends to thrive and grow — avoiding costs, generating new revenue streams, and preparing for a tougher regulatory environment and greater public scrutiny.[58] Hence, China and ASEAN countries are facing the challenges to adjust their economic growth models and reshape resource use in the economy. Such changes require overcoming enormous vested interests, deeply entrenched practices, and above all habits of thought and assumption.

RESPONSES TO ENERGY SECURITY CHALLENGES

Policies and Legal Framework

In China and ASEAN countries, there is greater awareness today of potential climate impacts on the regions, including reduced crop yields, waterstress, and extreme weather. There is also recognition of climate-related resource constraints on the economic growth, and that international economic structures and trading conditions have been shifting in response to these constraints. Thus many scientific and environmental elites, especially in China, support low carbon growth, not least because it can help lower high international energy prices. Others suggest that low carbon economic development provides the rational for upgrading a country's industries. It needs to successfully set the right goals and implement the right policies for low carbon economic development.

In China, there have been strong indications that the Chinese government sees the importance of sustainability and climate-friendly production as a critical component of future competitiveness. For example, the National Coordination Committee on Climate Change established in the late 1990s was upgraded in 2007 into a twenty-ministry National Leading Group to Addressing Climate Change. In 2006, former President Hu Jingtao approved The National Medium- and Long-Term Program for Science and Technology Development (2006–20), with a particular focus on developing renewable energy industries. Later, in March 2011 the government adopted its Twelfth Five-Year Plan (2011–15), which focused on seven strategically important spheres: energy-saving and environmental protection, next generation information technology, biotechnology, advanced equipment manufacturing, new energy (solar, wind, and biomass power), new materials and new-energy vehicles. The plan calls for non-fossil energy to meet 11.4 per cent of China's energy needs by 2015, and 15 per cent by 2020. To achieve that, Beijing is to spend US$1.7 trillion between 2011–15, in the form of investment, assistance for state-owned enterprises, and bank loans. That commitment was also reflected in the landmark climate deal announced during President Obama's visit to Beijing in November 2014. President Xi Jinping promised China would get 20 per cent of its energy from non-fossil fuel sources by 2020, and said its carbon emissions would peak by that year.[59]

Resource Efficiency

While China and ASEAN countries have certainly experienced substantial growth in demand for energy, they have tended to meet that demand inefficiently. One of the central aspects of any attempt to improve energy security and climate change is to adhere to best practices in improving efficiency in energy use and reducing energy intensity of economic growth.

China

China has been pursuing an aggressive energy conservation policy. Between 1980 and 2002, China's energy intensity (total primary energy consumption per dollar of GDP) improved at an average rate of 5 per cent per year. These efficiency gains were driven by domestic policies and programmes that incentivized greater efficiency. But this efficiency trend started to reverse itself between 2002 and 2005 due to the economic growth that came with joining the WTO, the subsequent increase in trade that drove strong industrial growth, and the growth in demands for energy.

In 2005, China regained its focus on promoting energy efficiency and launched mandatory energy intensity reduction goals as part of the Eleventh Five-Year Plan. The target was to reduce energy intensity by 20 per cent from 2005 levels by 2010. On 26 November 2009, China further announced that it will lower its carbon emissions relative to the size of its economy by as much as 45 per cent by 2020.[60]

Many of the successful programmes of the last five-year plan have been continued, expanded, or complemented in the Twelfth Five-Year Plan (2011–15). The Twelfth Five-Year Plan, adopted in 2011, includes targets to reduce energy intensity by 16 per cent and cut CO_2 intensity by 17 per cent by 2015, compared with 2010. Measures to achieve these targets have been reinforced by detailed industrial targets, particularly in terms of diversifying the power sector, which accounts for 55 per cent of China's coal use. There has been increasing activity and capital input in energy efficiency in China as a result of government's programmes to promote energy efficiency. From example, China's investment in energy efficiency in 2011 reached US$31 billion, 85 per cent of which was undertaken by companies.

Energy efficiency efforts over the past five years were principally aimed at reversing the trend of increasing energy intensity experienced in the early 2000s. The energy intensity goal for this plan is slightly scaled back to a 16 per cent reduction, and the priority is to lay the groundwork for future energy intensity improvements through economic transformation and introducing new and advanced technologies.

The central government also launched various programmes aimed specially at environmental protection. The total output of China's energy-saving and environmental protection industries reached 1.7 trillion *yuan* in 2009, accounting for about 5 per cent of China's GDP. Of the total, the environmental protection industry took up 26 per cent, the energy-saving industry 14 per cent and the resources recycling industry 60 per cent.[61]

Other policies, including stringent vehicle fuel economy standards and a new car-tax regime, that penalizes large cars, were introduced in 2006. Ordinary people in China have also been called to help save energy. In August 2007, the central government launched the all actions to implement the comprehensive work plan for energy conservation and pollutant discharge reduction campaign to emphasis the importance of lifestyle change, education, and adoption of new technologies.

ASEAN Countries

In order to achieve high energy efficiency targets, ASEAN countries have adopted related policies and measures. Indonesia is working toward reducing energy subsidies on a gradual basis and ensuring that subsidies are available only to low-income earners and small-scale industries. In October 2005, the government raised subsidized petroleum prices by 125 per cent in order to dampen demand and increase conservation. In May 2008, the government further increased petrol and diesel prices by nearly 30 per cent, and then in July 2008 raised liquefied petroleum gas (LPG) prices by 23 per cent. From May 2008 onwards, the government ceased paying subsidies to larger industrial electricity consumers.

Malaysia is currently in the process of formulating a comprehensive National Energy plan, which focuses on intensifying energy-efficiency initiatives in order to achieve more productive and prudent use of its remaining reserves. In 2008, Malaysia government introduced a broad package of reforms to energy subsidies. The package included subsidy reductions, cash rebates, windfall taxation on certain sectors and an expansion of the social safety net. Malaysian retail petrol prices were increased by more than 40 per cent in July 2008 and the following month the price of gas for power generation was raised by 124 per cent in Malaysia.[62]

The Thailand Power Development Plan 2010–30, released by the Ministry of Energy of Thailand in June 2012, targets a 25 per cent reduction of energy intensity of the country within twenty years (2011–30), resulting in the decrease of the country's power demand for economic growth. It is also targeting on increasing the share of renewable energy and alternative energy use by 25 per cent instead of fossil fuels within the next ten years, resulting in the replacement of some planned conventional (fossil fuels as coal-fired or gas-fired based) power plants by renewable power plants.[63]

In the Philippines, with the Philippines Energy Efficiency Project (PEEP), the government is aiming to reduce peak demand for power, reduce oil imports, and acquire revenue from the Clean Development Mechanism under the United Nations Framework Convention on Climate Change. The PEEP action plan includes measures such as replacement of incandescent light bulbs, the implementation of energy-efficiency standards for buildings and utilities, a ban on the importation of inefficient second-hand vehicles, and the establishment of an energy-efficiency and conservation testing centre.

Resource Diversification

China and ASEAN countries have been accelerating diversification of energy resources in order to reduce coal and oil dependence. Wider use of renewables including solar and wind power, nuclear power, hydro, and biomass fuels has been developing fast, especially in China. According to the planning of the National Development and Reform Commission, by 2020, renewable energy resources will satisfy 15 per cent of China's primary energy consumption, compared with about 7 per cent in 2005, and nuclear power will play an important role.[64] This will no doubt change China's energy structure greatly.

Nuclear Power

The Fukushima Daiichi nuclear power plant accident has directly affected nuclear energy policies not only in Japan but also in other countries including some European nations, prompting them to shift away from heavy dependence on nuclear energy. But the United States, France, Russia, and South Korea that have proactively promoted nuclear power generation, as well as emerging countries like China and India, have made no change to their respective nuclear promotion policies aimed at securing stable energy supply, preventing global warming, and maintaining and enhancing international competitiveness through nuclear industry development.

Currently China, Russia, and India lead the pack in nuclear power plant construction, with China accounting for more than a third of the nuclear power reactors currently built worldwide. Growth in nuclear capacity in China and its consequent export of nuclear technologies has created opportunities and challenges for both the country and the global nuclear regime. ASEAN countries also consider nuclear power as an important alternative source for electricity generation although there are still debates on the nuclear option in Southeast Asia.

China

Despite its late arrival in the nuclear field, China has the fastest nuclear industry in the world. At present, China has 17 units in operation situated in seven power plants in Zhejiang and Guangdong provinces, with 30 under construction and another 35 coastal plants being planned. Altogether, these facilities in operation yield slightly more than 10 gigawatts (GW) of total generating capacity, which amounts to only some 2 per cent of China's electricity needs.[65] China also has plans to expand domestic uranium enrichment, fuel fabrication, and reprocessing facilities to increase self-sufficiency in its closed fuel cycle.

China is a relative newcomer to nuclear power. It was not until late 1991 that China's first civilian nuclear power reactor went into operation at the Qinshan Nuclear Power Plant in east China's Zhejiang Province. It has moved quickly to indigenize nuclear technology, but the majority of the Chinese plants under construction now incorporate Generation II technology. The Fukushima accident in 2011 had prompted the Chinese government to raise the safety standards and slowed the pace of new construction. The Chinese State Council ordered safety reviews at all nuclear facilities, strengthened the safety management of operational facilities, ordered comprehensive reviews of all nuclear facilities under construction, and suspended the approval of new nuclear projects.

Furthermore, safety reviews that were completed by the end of 2011 concluded that all operating nuclear power plants and those under construction must meet higher standards. In October 2012, the State Council revised its projections of Chinese nuclear capacity. Instead of 70 GW by 2020, the new target is 58 GW, with 30 GW under construction. Inland nuclear power plant construction has been suspended until 2015, and all new nuclear powers must have Generation III standards.[66]

China has strictly implemented non-proliferation controls. Its nuclear export policy is guided by three principles: technology and equipment should be used for peaceful uses only, should be under appropriate IAEA (International Atomic Energy Agency) safeguards, and should not be transferred to third-party countries without Chinese consent. China has agreed not to export nuclear technology to countries that have been embargoed by the United Nations Security Council. Over the years, China has enhanced its export controls by joining the Zangger Committee in 1997, signing the IAEA Additional Protocol in 1998, and joining the Nuclear Suppliers Group in 2004. It has signed the Container Security Initiative with the United States and the Joint Declaration on Non-proliferation Security and Arms Control with the European Union, but it has not joined the Proliferation Security Initiative. China has also established an Export Licensing Catalogue of Sensitive Items and Technologies, and it still retains control for dual-use items that are not on its control lists. Finally, China conducts a number of training and consultation programmes to ensure that officials responsible for executing export control policy are properly prepared.

So far, China has exported power reactors only to Pakistan, and has no obvious national vision for exporting reactors, but there are a few signs of future interest: China's State Nuclear Power Technology Corporation has formed a joint venture with Westinghouse to market Westinghouse-based reactors, and along with China General Nuclear Corp (CGNC) and China National Nuclear Corp (CNNC), sold Chinese reactors to South Africa in 2013. In 2012, CGNC and CNNC were involved in a US$25 billion deal to build nuclear power plants in the United Kingdom.[67] Although Chinese involvement in the UK deal is limited to financing, there is no doubt that, like its high-speed train technologies, China will be making future efforts to export its nuclear technologies to its neighbouring countries.

ASEAN

In Southeast Asia, there are currently no nuclear power plants, but Vietnam, the Philippines, and Thailand have each included the introduction of nuclear power in their medium- and long-term power development plans, while Malaysia is looking at the possible deployment of nuclear power post-2020. Indonesia had plans for the introduction of a significant nuclear programme but these were cancelled in July 2009 due to tight credit and growing opposition.[68]

In 2011, Vietnam produced a Master Plan for National Power Development 2011–20 with the vision to 2030 which stipulates that the country will begin tapping nuclear energy by 2020. Currently, Vietnam has been proceeding with two nuclear plant construction plans in cooperation with Russia and Japan at Phuoc Dinh and Vinh Hain in Ninh Thuan province.[69] These plans are supported by generous terms provided by the governments of South Korea, Japan, China, Russia, and France, which would provide the technology and finance. After signing an Agreement between Vietnam and the United States in October 2013, U.S. Secretary of State John Kerry said, "Vietnam has the second-largest market, after China, for nuclear power in East Asia, and our companies can now compete. What is a US$10 billion market today is expected to grow into a US$50 billion market by the year 2030."[70] However, in January 2014, Prime Minister Nguyen Tan Dung announced that the construction of nuclear plants could be delayed until 2020, citing advice from the IAEA that "Vietnam must adopt a strict and effective solution for the nuclear power development."[71] The government wanted to ensure that all accuracy and safety aspects of the programme are addressed properly given that Vietnam is still developing its human resources in the nuclear field.

In Malaysia, Prime Minister Najib Razak announced a plan in 2009 to have a small-scale nuclear reactor.[72] Nuclear energy development is subsequently mentioned in the New Energy Policy 2011–14 (The Economic Planning Unit, 2010) but without a projected percentage of its total energy mix, as nuclear energy has always received strong public opposition in Malaysia. Civil society has expressed their objections to the nuclear option in a number of forums, illustrating public concern over nuclear waste. The Malaysian government, however, does not completely rule out the nuclear option. Its insistence in pursuing nuclear energy is evidenced in a statement made by Minister in the Prime Minister's Department Datuk Mah Siew Keong in early July 2014 which stated that the government will conduct a feasibility study aimed at building a nuclear power plant in ten years' time and carry out a comprehensive study including public acceptance and input from experts and non-government organizations.[73]

Since 1956, Indonesia has entertained the idea of nuclear power in anticipation of its future energy needs. Extensive preparations in accordance to IAEA guidelines and standards have been performed by the National Nuclear Energy Agency and the Nuclear Energy Regulatory Agency. But public opposition remains a critical factor that hinders Indonesian nuclear power plant development. In the early 1990s, feasibility studies identified suitable plant sites but strong civil society opposition stalled the plans.[74] Views of objections include distrust towards the government's capability in dealing with nuclear emergencies, financial constraints, Indonesia's vulnerability to nature disasters, and corrupt practices. In 2007, strong public opposition to nuclear power development led to President Yudhoyono's delayed approval in the establishment of a nuclear task force proposed by the Ministry of Energy and Mineral Resources.[75] The government's stand on nuclear energy is reflected in the National Energy Policy, passed in the parliament in January 2014, which also places the nuclear option as a last resort.

The government's low profile for nuclear energy is in stark contrast with national long-term planning that stipulates the nuclear energy share in Indonesia's energy mix by 2025, as nuclear energy is projected to make up 1.2 per cent and 1.7 per cent of Indonesia's energy mix in 2020 and 2025 respectively.[76] So despite opposition, relevant Indonesia officials claim that Indonesia has the necessary infrastructure and is ready for nuclear power plants although a decision is not expected until 2016.[77]

Like the rest of the world, there was a fundamental rethinking and debate on nuclear power development in East Asia following the Fukushima tsunami and destruction of its nuclear power plants in March 2011. The safety risks of nuclear power led to a concern and major shift in perceptions in many Southeast Asian countries. However, Southeast Asian governments have not abandoned the idea. They have reassessed and want to proceed with their original plans. This is because they are keen to reduce their reliance on imported energy. The search for energy security is the main factor for turning to nuclear energy, as energy security largely means energy independence. While the cost of nuclear power remains high, the rapidly increasing exploitation of shale gas will drive down the region's energy costs. Clean coal technologies pioneered by

China and the United States will also reduce the carbon emissions of coal-fired power plants, although there will be a time lag before widespread adoption occurs. However, concerns about energy security help to drive policymakers to search for energy independence.

There is, of cause, considerable uncertainty about the prospects for nuclear power in Southeast Asia as there are many challenges to overcome, including public acceptance, financing, site selection, long-term storage of spent fuel, developing safety and security regulations, and the building up of human resources and technological capabilities. Although advocates of nuclear power technology, especially the exporters of nuclear power plants, argue that the technology used will be more advanced than in the Fukushima reactors, they have neglected Southeast Asia's bureaucratic culture of obedience and deference as well as their management levels which could affect the quality and safety.

Regional Energy Cooperation

In working to enhance energy security and slow climate change, China and ASEAN countries have advocated a variety of measures and participated actively in regional energy cooperation. These include promoting the development and use of natural gas and renewable energy; spreading clean-coal technology and energy conservation practices; and transferring know-how on human resource development and institutional design. Such cooperation efforts have been implemented not only on an ASEAN basis but also in multilateral frameworks such as the ASEAN+3 (ASEAN with China, Japan, and South Korea) and East Asia Summit.

ASEAN

Energy cooperation within ASEAN had begun as early as in the 1970s, and is viewed as part of its broad goal for regional economic integration, including a trans-ASEAN gas pipeline and trans-ASEAN power grid. Intra-ASEAN energy cooperation is led by annual meetings of ministers supported by the Senior Officials Meeting on Energy process, the ASEAN Secretariat, and the ASEAN Center for Energy (ACE). Although the ASEAN Plan of Action for Energy Cooperation 2010–15 emphasizes long-term efforts to promote energy security and clean energy, the document does not mention emergency preparedness or oil stocks. This is due to the fact that ASEAN energy collaboration previously had a strong focus on oil supply security in the 1970s and 1980s. For example, the ASEAN

Council on Petroleum (ASCOPE) was established in 1975 as a coordinating mechanism with industry, and in 1977 ASCOPE established a petroleum-sharing scheme to be implemented by member countries' state-owned oil companies. Since ASEAN included both importers and exporters, the sharing plan was devised to manage periods of both shortage and oversupply in exports of 20 per cent.

Today, talks of regional energy cooperation is buoyant in Southeast Asia. These are promulgated in the five-year ASEAN Plans of Action for Energy Cooperation (APAEC) which are researched and drafted by the ACE based in Jakarta. The current plan, APAEC 2010–15, includes seven programme areas: the ASEAN power grid, the trans-ASEAN gas pipeline, coal and clean coal technology, renewable energy, energy efficiency and conservation, regional energy policy and planning, and civilian nuclear energy.[78]

In January 2013, a project on ASEAN Energy Market Integration (AEMI) was conceived and developed at Chulalongkorn University, Thailand. The AEMI initiative was fuelled by an emerging consensus among a number of ASEAN academics that a successful AEMI would be a necessary condition for achieving sustainable growth in the Framework of AEC. It would enhance energy security and environmental viability across the region and undoubtedly yield significant benefits for all involved, from the economic, social and environmental perspectives. The ultimate objective of the AEMI Group is the adoption of AEMI within the framework of AEC, and its deployment through 2030.

ASEAN itself has announced plans for an integrated network of electricity transmission and distribution lines (the ASEAN Power Grid) and a separate interconnected network of natural gas pipelines (the Trans-ASEAN Gas Pipeline). Regional cooperation in the form of interstate electricity distribution has occurred in the Greater Mekong Subregion between Thailand, Laos, Cambodia, China, and Vietnam, and bilateral trading of natural gas has occurred via pipelines between Indonesia, Malaysia, Singapore, and Thailand.

However, ASEAN faces many challenges in developing these energy cooperation programmes. These intensifying energy ties will not only require extensive investment in producer countries, but also entail expensive and strategically vulnerable infrastructure, including pipelines, hydro-power dams, deep-sea ports, oil and gas terminals, and storage facilities. Thus, ASEAN needs to seek cooperation with its neighbours in Northeast Asia and beyond.

ASEAN+3

Established in 1996, the ASEAN+3 arrangement with China, Japan, and South Korea proved to be a useful forum during the Asian financial crisis and has since become a positive factor in strengthening ASEAN's capabilities for managing energy emergencies. The grouping's contributions include joint meetings and studies on oil stocks as well as the establishment in 2007 of an energy security system.[79]

This cooperation had its origins in 2003 when Asian concerns over energy supply security heightened in anticipation of the Iraq War. These concerns led to the first ASEAN+3 energy ministerial meeting, which was held in Manila in June 2004. The importance of oil stockpiles was recognized in the ministerial statement, and during the meeting the Philippines suggested creating a regional stockpile at the former U.S. naval base at Subic Bay. In August 2008 at their fifth meeting in Bangkok, the ASEAN+3 energy ministers announced the development of the ASEAN+3 "oil stockpiling road map", which could be based on four principles: voluntary and non-binding agreements; mutual benefits; mutual respect, including respect for bilateral and regional cooperation; and a step-by-step approach with long-term perspectives.

A working group was formed to develop the road map, and two years later at a June 2010 IEA workshop in Jakarta on emergency planning, a representative from the ACE made a presentation on the initiative. The presentation concluded that due to differences in the economic situations of various ASEAN+3 countries, it was not presently possible to develop a detailed road map with concrete targets for all members.[80]

Despite this setback, the ninth ASEAN+3 energy ministerial meeting in Cambodia in September 2012 authorized continued work on the road map, including proposed workshops on planning and construction of oil stockpiles, and encouraged stakeholder cooperation and continued cooperation with the IEA. These same themes were discussed again in September 2013 at the grouping's tenth meeting in Indonesia. More recently, the ASEAN+3 group Oil Stockpiling Roadmap Workshop held in February 2014 in Siem Riep, Cambodia, included presentations on the economic benefits of investing in a join stockpile.

East Asia Summit

As a leader-led forum that focuses on broad political, economic, and strategic issues, the first East Asia Summit (EAS) meeting was held in Malaysia in December 2005, at which time EAS leaders agreed to enhance their cooperation by promoting energy security.

At the second summit in the Philippines in January 2007, the leaders held a special session on energy, which resulted in the Cebu Declaration on East Asian Energy Security. The Cebu Declaration reaffirmed the "collective commitment to ensuring energy security for our region" and addressed a wide range of energy issues including the goal to "explore possible modes of strategic fuel stockpiling such as individual programmes, multi-country and/or regional voluntary and commercial arrangements". It recognized that less developed members would need assistance in their national capacity-building to achieve these energy security goals and added that "the necessary follow-up actions to ensure implementation will be undertaken through existing ASEAN mechanisms in close consultations among EAS participants".[81]

The leader also agreed to establish the EAS Energy Cooperation Task Force based on ASEAN mechanisms in order to carry out the summit's energy work. In addition, they endorsed Japan's proposal for the establishment of the Economic Research Institute for ASEAN and East Asia (ERIA) to serve as the de facto secretariat on energy issues. Also at Cebu, Singapore proposed that the EAS energy ministers meet, and the first EAS energy ministerial meeting was held a few months later in Singapore in August 2007. Since then, it has become a standard practice for EAS energy meetings to be held in conjunction with ASEAN and ASEAN+3 meetings. The Cebu Declaration notwithstanding, oil stocks and energy-contingency planning have not yet been a focus for EAS energy ministers.

The first EAS energy ministerial statement was silent on the issue when it established three work groups for energy cooperation under the EAS umbrella: energy efficiency and conservation, energy market integration, and biofuels. In November 2007, EAS leaders issued the Singapore Declaration on Climate Change, Energy and the Environment, which reflected their focus on long-term measures to enhance energy security and emphasized actions to promote clean energy and efficiency in order to help combat climate change.

The second EAS energy ministerial in Bangkok in 2008 expressed concern over high oil prices and ministers "affirmed to vigorously take their actions in such areas as enhancing emergency preparedness".[82] The third EAS energy ministerial in Myanmar in 2009 subsequently addressed the essential issue of data transparency by highlighting the importance of the Joint Oil Data Initiative. The EAS grew to eighteen nations in November 2011 when the United States and Russia joined, and at the sixth meeting in Cambodia in 2012, it was agreed that the ERIA would work jointly with the IEA on energy outlook studies. Finally, at the seventh meeting in Indonesia in September 2013, the ministers added a new work stream on renewable and clean energy.

In summary, energy cooperation between ASEAN and East Asian countries presents great opportunities and advantages. In East Asia, Japan, South Korea, and to a large extent China are leading countries in energy development and overseas investment. Thus, they can cooperate with ASEAN countries in improving energy technologies and exploring energy resources. Indonesia is particularly important to the East Asia region because of its endowment of energy resources, including coal, natural gas, and other mineral resources, which can help secure the energy supplies of Asian countries. Brunei, Malaysia, Myanmar, and Vietnam also have large potential reserves of oil and gas and could benefit from technology transfer and investment from China, Japan, Korea, India Australia, and New Zealand (ASEAN+6 grouping). The refinery capacities in South Korea and Singapore and the oil storage capacity in Japan could provide further benefits with more integration.

Notes

1. IMF, *World Economic Outlook Database*, October 2013.
2. Ibid.
3. Ibid.
4. IEA, *World Energy Outlook Special Report 2013: Southeast Asia Energy Outlook*, p. 34.
5. Joseph Y.S. Cheng, "A Chinese View of China's Energy Security", *Journal of Contemporary China* 17, no. 55 (2005): 297–317.
6. ADB, *Key Indicators for Asia and the Pacific 2012.*
7. IEA, *World Energy Outlook 2012.*
8. IEA, *Southeast Asia Energy Outlook 2013*, p. 39.
9. Li Qiang and Qi Xingkang, "World Energy Structure and Choices of Chinese Energy Strategy", *Procedia Earth and Planetary Science* 1, no. 99 (2009): 1723–29.
10. Xu Xiaojie, *Energy Black Swan: Global Games and Chinese Options* (Beijing: Shehui Kexue Chubanshe, 2012).
11. John Ryan and Clara Gillispie, "New Outlooks for Asian Energy Security", *NBR Workshop Report*, May 2014, <http://www.nbr.org/research/activity.aspx?id=448> (accessed 13 May 2014).

12. "Woguo 2020 nian zhanlue shiyou chubei liang jiang da shijie dier" [China's Strategic Oil Reserves Will Reach the Second Largest in the World], <http://www.china5e.com/show.php?contentid=207204> (accessed 16 September 2012).
13. Ibid.
14. "Woguo 2020 nian zhanlue shiyou chubei liang jiang da shijie dier" [China's Strategic Oil Reserves Will Reach the Second Largest in the World], 30 January 2012, <http://www.china5e.com/show.php?contentid=207204> (accessed 16 September 2012).
15. IEA, *World Energy Outlook 2009*, p. 607.
16. Shinji Omoteyama, "Energy Sector Situation in Vietnam", Institute of Energy Economics, Japan, May 2009.
17. Ibid.
18. "Woguo chenwei disheng da tiangnan qi xiaofei guo" [China Became the Third Largest Gas Consuming Country], *China Petroleum Daily*, 20 May 2014.
19. Xu Yongfa, president of CNPC Economics and Technology Research Institute, as quoted in Interfax, *China Energy Weekly*, 27 February–2 March 2012, p. 8.
20. Based on the CNPC figures of 550 bcm of gas demand in 2030 and figures from IEA.
21. EIA, "World shale oil and shale gas resource assessment", June 2013, <http://www.advres.com/pdf/A_EIA_ARI_2013%20World%20Shale%20Gas%20and%20Shale%20Oil%20Resource%20Assessment.pdf> (accessed 26 March 2015).
22. Energy Information Administration (EIA), "World Oil Transit Chokepoints", *Country Analysis Briefs*, January 2008, <www.eia.doe.gov> (accessed 16 September 2012).
23. The first Joint Maritime Strategy document of the U.S. Navy, Marine Corps and Coast Guard promulgated in October 2007, "A Cooperative Strategy for 21st Century Sea Power", clearly states this. This is an important geographic shift of focus for the U.S. Maritime Military Power from the Atlantic and further drives home the point of overall shift of power to Asia. See Ashok Sawhney, "Indian Naval Effectiveness for National Growth", *RSIS Working Paper*, No. 197, 7 May 2010.
24. As North and Borschberg note elsewhere in this issue, the 500-mile-long Straits of Malacca is the main link between the Indian Ocean and the South China Sea, with the main sea lanes for tankers inbound from the Middle East through the Malacca and Singapore Straits. An average of twenty-six such ships transit the Singapore Strait on a daily basis. Because of the relative shallowness of the Strait, there are limitations on the size of tankers which can transit it.

25. Located between Java and Sumatra, the Sunda Strait is 50 miles long and is another alternative to the Straits of Malacca. Its north-eastern entrance is 15 miles wide; because of its strong currents and limited depth, deep-draft ships of over 100,000 deadweight tons do not transit the strait, thus it is not heavily used.
26. The Lombok Strait in Indonesia is wider, deeper, and less congested than the Straits of Malacca. It separates the islands of Lombok and Bali. A ship's route would often take it through the Makassar Strait between Borneo and Sulawesi.
27. IEA, *World Energy Outlook 2009*, p. 607.
28. Marc Lanteigne, "China's Maritime Security and the 'Malacca Dilemma'", *Asian Security* 4, no. 2 (2008).
29. Shi Chunlin, "The Impact of United States on the Safety of China's Pacific Shipping Routes and Countermeasures to be Taken", *China Maritime Safety*, no. 2 (2011).
30. Peper Escobar, "So Many Secrets in the East China Sea", *Asia Times Online*, 11 December 2013.
31. Philip Andrews-Speed and Roland Dannreuther, *China, Oil and Global Politics* (Abingdon: Routledge, 2011), p. 140.
32. Jean A. Garrison, *China and The Energy Equation in Asia* (Colorado: First Forum Press, 2009), p. 93.
33. Jennifer L. Turner, "China's Green Energy and Environmental Policies", testimony given to U.S.-China Economic and Security Review Commission, 8 April 2010, <http://www.chinafaqs.org/files/chinainfo/jennifer_turner_testimony_4-8-10.pdf> (accessed 26 March 2015).
34. IEA, *World Energy Report 2009*.
35. Institute of Energy Economics, Japan, *Asia/World Energy Outlook 2012*, <http://eneken.ieej.or.jp/data/4683.pdf> (accessed 26 March 2015).
36. IEA, *World Energy Report 2013*, p. 45.
37. Ibid., p. 487.
38. IEA, *World Energy Outlook 2009*.
39. Gordon White, *Riding the Tiger: The Politics of Economic Reform in Post-Mao China* (London: Macmillan, 1993), p. 4.
40. David Martin Jones and M.L.R. Smith, *ASEAN and East Asian International Relations* (Cheltenham: Edward Elgar Publishing, 2006), p. 84.
41. John Lee, "China's Geostrategic Search for Oil", *The Washington Quarterly* 55, no. 3 (2012).
42. Ibid.
43. "Global 500 for 2012", *Fortune Magazine*, <http://topforeignstocks.com/2013/03/13/the-fortune-global-500-companies-list-2012/> (accessed 26 March 2015).
44. John Lee, "China's Geostrategic Search for Oil", *The Washington Quarterly* 55, no. 3 (2012).

45. "China's Premier Wen Jiabao Targets Social Stability", BBC, 5 March 2011, <http://www.bbc.co.uk/news/world-asia-pacific-12654931> (accessed 26 March 2015).
46. David Martin Jones and M.L.R. Smith, *ASEAN and East Asian International Relations* (Cheltenham: Edward Elgar Publishing, 2006), p. 89.
47. Benjamin K. Sovacool and Vu Minh Khuong, "Energy Security and Competition in Asia: Challenges and Prospects for China and Southeast Asia", in *ASEAN Industries and the Challenge from China*, edited by Darryl S.L Jarvis and Anthony Welch (Basingstoke: Palgrave Macmillan, 2011).
48. The current Indonesian government under President Joko Widodo (Jokowi) is reducing the level of subsidies by removing fuel for private vehicles from the scheme and eliminating fuel subsidies altogether.
49. Cut Dian R.D. Agustina, "The Regional Effects of Indonesia's Oil and Gas Policy: Options for Reform", *Bulletin of Indonesian Economic Studies* 48, no. 3 (2012): 369–97.
50. IEA, *Southeast Asia Energy Outlook 2013*, p. 23.
51. "The Green-Growth Twofer", *The Economist*, 10 January 2014.
52. Ibid.
53. IEA, *Southeast Asia Energy Outlook 2013*, p. 23.
54. Ibid.
55. *IMF Country Report No. 13/362*, December 2013, p. 57.
56. *IMF Country Report No. 12/278*, September 2012, p. 10.
57. Keoni Indrabayu Marzuki, "Balancing the Short and Long Term in Indonesia Fuel Subsidy Debate", *East Asia Forum*, 21 October 2014.
58. World Economic Forum (2012), "More with Less: Scaling Sustainable Consumption and Resource Efficiency", <https://www.cdp.net/en-US/News/Documents/more-with-less.pdf> (accessed 26 March 2015).
59. "Asia Pushes Hard for Clean Energy", *International New York Times*, 18 November 2014.
60. "China Sets Target for Emission Cuts", *The Washington Post*, 27 November 2009.
61. "Fast Growth for Environmental Protection Market: Report", *China Daily*, 7 May 2010.
62. IEA, *World Energy Outlook 2009*, p. 604.
63. Ministry of Energy of Thailand, "The 20-Year Energy Efficiency Development Plan 2011–2030", <http://gec.jp/gec/jp/Activities/fs_newmex/2012/2012_mrvds12_jSmartenergy_thailand_ann.pdf> (accessed 26 March 2015).
64. Joseph Y.S. Cheng, "A Chinese View of China's Energy Security", *Journal of Contemporary China* 17, no. 55 (2005): 297–317.
65. Richard Weitz, "Beijing Confronts Japanese Nuclear Meltdown", *China Brief* xi, issue 6 (8 April 2011).

66. "New Nuclear Suppliers", Report of Workshops Hosted by the CSIS Proliferation Prevention Program, <http://csis.org/files/publication/131211_ppp_nuclear_suppliers.pdf> (accessed 26 March 2015).
67. Ibid.
68. IEA, *World Energy Outlook 2009*, p. 558.
69. IEEJ, *Asia/World Energy Outlook 2012*, p. 17.
70. "Agreement Opens U.S.-Vietnam Nuclear Trade", *World Nuclear News*, 10 October 2013, <https://nuclearstreet.com/pro_nuclear_power_blogs/b/world_nuclear_news/archive/2013/10/10/agreement-opens-us-vietnam-nuclear-trade.aspx#.VAa30lI6zK8> (accessed 3 September 2014).
71. "Vietnam: A Delay in Nuclear Power Raises Energy Security Concerns", *Manila Times*, 26 January 2014.
72. Bernama, "Malaysia Keen to Develop Small-Scale Nuclear Reactor: Najib", *Education News Update*, 2 June 2009, <http://web3.bernama.com/education/newsr.php?id=415373> (accessed 26 March 2015).
73. Bernama, "Government to Conduct Feasibility Study on Building Nuclear Plant: Mah", *News Straits Times*, 7 July 2014.
74. S. Amir, "Nuclear Revival in Post-Suharto Indonesia", *Asian Survey* 50 (2010): 265–86.
75. Ibid.
76. Biofuel is predicted to form 4.5 per cent and 5.1 per cent of total energy mix in 2020 and 2025, while geothermal is projected to make up 5.1 per cent and 5.2 per cent (The Ministry of Energy and Mineral Resources Republic of Indonesia, 2006).
77. IAEA, "Country Nuclear Power Profiles 2013 Edition, Indonesia", <http://www-pub.iaea.org/MTCD/Publications/PDF/CNPP2013_CD/countryprofiles/Indonesia/Indonesia.htm> (accessed 26 March 2015).
78. ASEAN Secretariat, "ASEAN Plan of Action for Energy Cooperation 2010–2015 Bringing Policies to Actions: Towards A Cleaner, More Efficient and Substainable ASEAN Energy Community", <http://cil.nus.edu.sg/rp/pdf/2010%20ASEAN%20Plan%20of%20Action%20on%20Energy%20Cooperation%20%28APAEC%29%202010-2015-pdf.pdf> (accessed 19 August 2014).
79. The ASEAN+3 energy system comprises a communication system and a database. The communication system consists of a real-time, web-based chat room with displays limited to one hundred characters and a bulletin board with a two thousand kilobyte file size. See Beni Suryadi, "ASEAN+3 Oil Stockpiling Roadmap and ASEAN+3 Energy Security System", presented by the ACE at the IEA-APEC/ASEAN Emergency Response Exercise, Bangkok, 2–3 May 2011.
80. Beni Suryadi, "Development of Oil Stockpiling Roadmap for ASEAN+3", presented at a joint IEA and Indonesian Ministry of Energy and Mineral Resources workshop, Jakarta, 16–17 June 2010.

81. "Cebu Declaration on East Asian Energy Security", 15 January 2007, <http://www.asean.org/asean/external-relations/east-asia-summit-eas/item/cebu-declaration-on-east-asian-energy-security-cebu-philippines-15-january-2007-2> (accessed 26 March 2015).
82. "Joint Ministerial Statement of the Second East Asia Summit Energy Ministers Meeting", 7 August 2008, <http://www.asean.org/asean/external-relations/east-asia-summit-eas/item/joint-ministerial-statement-of-the-second-east-asian-summit-energy-ministers-meeting-bangkok-7-august-2008> (accessed 26 March 2015).

3

CHINA'S ENERGY QUEST IN SOUTHEAST ASIA

Southeast Asia lies at the junction of South Asia and East Asia, and is traditionally seen by the United States, China, Japan, India, and other big powers as their respective spheres of influence. China has been a keen player in this region for historical reasons in view of the existence of a large diaspora, energy resource supply, trade and investment linkages and protection of its maritime interests. Energy resource cooperation is an important part of China–ASEAN relations. It has been driven under the background of China's energy diversification strategy, the development of China–ASEAN FTA (Free Trade Agreement) and regional economic integration. As China expanded its foreign direct investment (FDI) to Southeast Asia after the global financial crisis in 2008, cooperation in this field developed to a new level, extending from energy trade to energy resource exploration and related infrastructure-building.

CHINA'S ENERGY DIVERSIFICATION STRATEGY

China imports much of its oil from the Persian Gulf region through the Strait of Hormuz, where security is dependent on the U.S. Navy. The most prominent issues in the Middle East include the threat posed by Iran's nuclear programme, the emergence of the Islamic State of Iraq and Syria (ISIS) as a destabilizing force in Syria and Iraq, continued instability in North Africa, and the acceleration of Israeli–Palestinian conflicts. These are conflicts involving a great percentage of the world's major energy suppliers. Beijing is concerned about the implications of a potential reduction of U.S. presence in this region. Therefore Chinese leaders are facing questions about what they can and should do to protect the security of energy supply.

China's Approaches to Energy Resource Security

"Mercantilist" or "Liberalist" Approaches?

In a globalized economy, securing the supply of energy resources is a pressing issue for many states. All countries rely on energy to power their electricity and transportation systems, and the emerging economies require mineral resources for their rising industrial sectors. But energy resource production and consumption centres are often located in different national spaces. For emerging economies like China and India that lack local reserves of energy resources, this poses a particular challenge for their energy security. The need to import natural resources exposes these economies to high levels of resource insecurity, and many governments adopt energy resource security strategies to mitigate these risks. According to Flynt Levrett, we can identify two contrasting types of energy resource security strategy, defined by whether they employ state- or market-based mechanisms — mercantilist and liberal resource security strategies.[1]

Mercantilist resource security strategies are a state-directed approach to quest energy resources abroad by economies that must import resources. They involve the use of economic and foreign policy instruments by national governments to help their state-owned national energy companies (NECs) secure access to overseas hydrocarbon resources on more privileged basis than simply supply contracts based on market prices. The strategy aims to ensure control over resource supplies by having national firms that own projects at the site of production.[2] Government policies are deployed to assist national firms invest in overseas resource or resource-related projects, including foreign policies designed to enhance bilateral links with key supplier states in an effort to improve the environment for such investments — known as "resource diplomacy".

In contrast, liberal resource security strategies are market-oriented, and involve a consumer government mainly relying upon international markets to address its energy resource security concerns. The strategy draws on liberal ideas to argue that properly-functioning international resource markets carry advantages associated with their efficiency, stability, and transparency.[3] Rather than pursuing national control of a few key suppliers, resource security is maximized by governments turning to open, competitive, and reliable international markets. In this model, state policies aim to integrate national energy industries with existing international markets, and to ensure that these markets provide a stable and transparent climate in which all consumers can purchase needed natural resources.[4]

China's overseas energy investment has drawn much interest from foreign observers and has raised different arguments over the nature of its strategy and geopolitical implications. These arguments focused on China's use of state-owned enterprises to "lock up" new sources of resource supply, its extension of financial support to subsidize its firms' investments abroad, and its bilateral resource diplomacy initiatives in many regions, aiming to smooth the way for such investment.[5] Leverett believes that China has taken the lead in developing a resource-mercantilist approach to energy security, with India effectively following China's example.[6] China's resource quest has also become a controversial issue for international politics, with the European and U.S. governments publicly arguing that China's strategy is not only mercantilist, but also provoking interstate competition resources.[7] These actions are regarded as security threats and sources of conflict.[8]

Over the past years, Chinese government has promoted several initiatives to address the energy resource security problem. Foremost among these has been the "going out" strategy, which called for Chinese energy companies to begin to make resource investment to help to increase China's energy security, as "national ownership of foreign resource suppliers is believed to act as a hedge against the risks associated with resource import-dependence, since firms are expected to preferentially direct the resources they produce back to their home market".[9] Thus China has become one of the most active participants in a global race for resources over the last decade, and has made energy security-oriented investments in many overseas oil, gas, copper, coal, bauxite, and iron ore projects. With the support of plentiful credits from Chinese state-owned banks, China's oil companies have expanded their investment activities far and wide, including those politically unstable areas such as Sudan, Venezuela, Iraq, Iran, Syria, and Myanmar. The fact that Beijing has seemingly little concern about political risks and inefficient investment reaffirms the reality that China's economic objectives remain largely subordinate to political consideration.[10]

These perceptions seem to derive from the phenomenon that China's overseas direct investment is most focused on energy resources, conducted by state and enterprises (SOE) and secured by government support. In reality, however, the story is less sinister but more complicated. One basic fact is that while the majority of Chinese energy and resource firms are still SOEs, their shares are decreasing with rapid market transition and decentralization occurring in China.

China's "going out" strategy was initiated in the late 1990s. It reflects and is an extension of China's domestic economic structures. The domestic structures of China's political economy (its heavily state-owned economy), especially in the energy sector, mean that it had to rely on SOEs during the initial period of its "going out" when its private companies lacked overseas experiences and strength. Moreover, the arguments of some Chinese analysts of not allowing the free market mechanism to dominate activities in "strategic sectors" had been reflected in Beijing's energy security policy. In Beijing's view, "strategic sectors are simply too important to be left to market forces alone"[11] and private companies. Hence, the SOE-led and politically-driven approach to energy security policy was prevailing during the initial period of China's "going out". As a latecomer to the international energy market, China also found that most of the good oil and gas assets in stable and respectable countries were unavailable because they were already owned by national companies in the host countries or by Western oil companies. Therefore, "China has been forced to turn to countries where the U.S. sanctions forbid American companies from doing business, such as the Sudan and Iran".[12]

Moreover, the investment climate in developed economies was often more complex and less welcoming. For example, the U.S. Congress' refusal to allow CNOOC (China National Offshore Oil Corporation) to take over the U.S.-based UNOCAL in 2005 on the excuse of safeguarding national security highlighted the sensitivity in many Western countries about China's form of state capitalism. When the Aluminum Corporation of China Ltd., another state-owned company, was similarly prevented from investing in Australia-based iron ore supplier Rio Tinto in 2009, it was clear that China's outward economic expansion was going to have to overcome many political hurdles if it was to play a role in China's resource security policy.[13] Thus, developing countries in particular were targeted by Chinese companies because they hoped to gain from Chinese aid and investment and were therefore generally receptive to the approach, and because there were generally few political obstacles for China to overcome.

Figure 3.1 shows that by 2010, 23 per cent of China's overseas equity oil production (the National Oil Companies or NOCs own or have controlling stakes in actual overseas oilfields) was in Kazakhstan, 15 per cent in both Sudan and Venezuela, 14 per cent in Angola, 5 per cent in Syria, 4 per cent in Russia, and 3 per cent in Tunisia. Nigeria, Indonesia, Peru, Ecuador, Oman, Columbia, Canada, Yemen, Cameroon, Gabon, Iraq,

FIGURE 3.1
Shares of China's Overseas Equity Oil Production, 2010

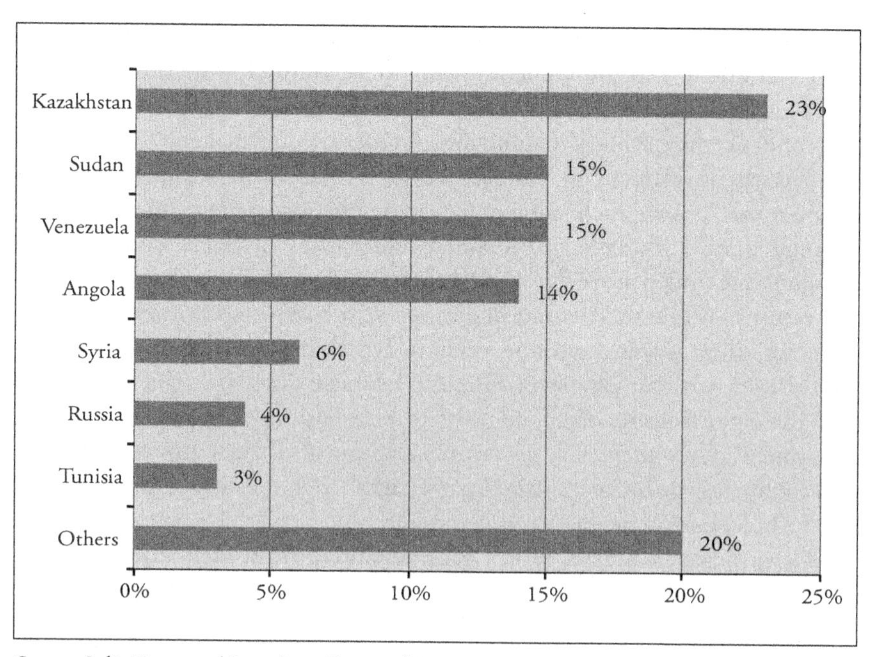

Source: Julie Jiang and Jonathan Sinton, "Overseas Investment by Chinese NOCs", IEA, February 2011, <http://www.iea.org/publications/freepublications/publication/overseas_china.pdf> (accessed 26 March 2015).

Azerbaijan, and Uzbekistan make up the remaining 20 per cent. Chinese NOCs now operate in over thirty countries and have equity production in at least twenty countries.

According to the International Energy Agency (IEA), Chinese overseas equity production amounts to around 28 per cent of total current Chinese importing requirements, which were 4.8 million barrels per day in 2010.[14] NOCs owned/controlled overseas sites are currently producing around 1.37 billion barrels per day, and it is predicted that Chinese NOCs' overseas equity production will reach around 2 million barrels per day by 2020, which is significantly less than the official 2020 target of 4 million.[15] Based on this equity investment approach, China's "energy-strategic areas" have been formed, including the North African area centred on Sudan, the Central Asian area centred on Kazakhstan, South American area centered on Venezuela, and the Middle East area centred on Iran.[16]

But as on whether or to what extent "equity oil" can ensure China's energy security and "act as a hedge against the risks associated with resource import-dependence", there exists different views and doubt. In reality, the interest of Chinese energy companies do not always accord with those of the government and are sometimes in direct conflict with them. Just as Eric Downs, former research fellow of American Brookings Institute, points out, one of the primary complaints Chinese policymakers make about the overseas investment of Chinese SOEs is that "each soldier is fighting his own war", with each company placing their corporate interest above national ones.[17] Actually, for Chinese enterprises, overseas commitment decisions are independently made based on their evaluation of the risks and returns. While they would like to secure whatever government support they can, they would continue their internationalization and investment abroad even without any such support. In most cases, it is the SOEs that find the opportunities first and initiate negotiations over the prospective investment. They then seek government approval of their investment plan and lobby for financial or diplomatic support if needed. Thus they are largely independent from the government.

Nevertheless, Chinese NOCs have benefitted from their experiences in some of those oil and gas producing countries which were sanctioned by Western countries. For example, from 2004 to the end of 2012, China's accumulated investment value in Sudan increased from US$0.17 billion to US$1.24 billion.[18] In 2006, about 50 per cent of China's equity oil came from Sudan, and 65 per cent of Sudan's oil exports went to China,[19] and in 2011, China's oil import from Sudan accounted for 5 per cent in its total oil imports.

In Iran, as Western countries have decreased their trade and investment there, China stepped in to fill the void and began to play a major role in Iran's oil industry. From 2004 to 2012, China's investment in Iran increased from US$0.05 billion to US$2.1 billion, most of which was in the oil and gas sectors. Since the beginning of the twenty-first century Iran has been consistently one of China's largest suppliers of crude oil, and in 2011 China's oil imports from Iran accounted for 11 per cent of its total oil imports, and its oil import from Middle East accounted for 45 per cent.[20] Moreover, China is the biggest trade partner of Iran. The volume of China–Iran bilateral trade steadily increased from US$5.6 billion in 2003 to US$45 billion in 2011. The political situation in these countries may have provided China with certain competitive advantages, but these carry substantial risks and can exact both financial and reputational costs, both for the NOCs and the Chinese government.[21]

While China's overseas investment soared, important changes have also taken place in the nature of the investments in recent years. Market forces have become major driving forces for Chinese firms, especially non-SOEs to invest overseas. A key motivation for Chinese companies investing overseas, overlooked by the detractors who view Chinese SOEs as tools of the state, is economic self-development. International exploration and production are a means of improving the technological, technical, and managerial capabilities of Chinese companies and facilitating the export of related facilities, technology, and labour. One dynamic changing phenomenon is that while the majority of Chinese energy firms are still SOEs, non-SOEs have been dominant as China's overseas investors in terms of numbers. Although the average size of their investment remains small compared to those by large SOEs, their share in China's total outward FDI has increased. For example, by the end of 2013, among US$544 billion non-financial outward FDI stock, state-owned enterprises had taken a share of 55 per cent, and that of the non-state enterprises had reached 45 per cent, increasing by 4.6 per cent compared with the previous year. Among the non-state enterprises, limited liability companies had taken a share of 31 per cent while the shares of incorporated companies, private companies, joint-stock cooperative enterprises, foreign-invested enterprises, Hong Kong, Macao and Taiwan-invested enterprises, collective enterprises and others accounted for 14 per cent (see Table 3.1).

TABLE 3.1
Structure of China's Outward FDI Stock

Type	Share (%)
Limited liability companies	55.2
Private enterprises	30.8
SOEs	2.2
Incorporated companies	7.5
Joint-stock enterprises	2.0
Foreign enterprises	3.0
Hong Kong, Macau and Taiwan-invested enterprises	0.4
Others	0.6
Total	100

Source: Ministry of Commerce of PRC, *2013 Statistical Bulletin of China's OFDI*, p. 106.

China Adjusting Its "Going Out" Strategy

The past years have seen Chinese oil companies face potential risks and rising costs in their overseas operation. Since 2010, frequent crisis and political changes have afflicted those countries that were considered as China's overseas energy-strategic areas. China suffered great potential loss in its large investments in Sudan, Iran, Syria, Libya, and Venezuela which were politically unstable countries. For example, Libya has the largest proven oil reserves in Africa. Before the civil war broke out in 2011, Libya's crude oil production reached 1.6 million barrels per day, of which about 120 million barrels were exported to European countries per year. Because of the outbreak of the war, China's imports of crude oil from Libya decreased dramatically from 7.37 million tons in 2010 to 2.59 million tons in 2011, a drop of 64.8 per cent.[22] China's imports of crude oil from Sudan (North and South Sudan) decreased from 13 million tons in 2011 to 2.5 million tons in 2012, a drop of 80 per cent.[23] A large amount of China's investment in energy facilities and infrastructures also got damaged.

Thus China's overseas projects have been facing potential constraints and rising costs. The Asia Pacific Energy Research Centre in Japan has created an index to compare the relative position of oil supply security in China, Japan, Korea, and the United States with respect to four factors: equity oil ratio, self-sufficiency ratio, independence from oil ratio, and political stability of crude oil import sources. The result shows that China's political stability index for crude oil import sources represents the lowest level among these four countries.[24] What adds to worries is that "With the Americans moving toward a greater degree of hemispheric energy security, led in no small part by greater levels of self-supply in the U.S., there have emerged growing levels of concern in China about further uncertainties in worldwide energy dynamics."[25]

Moreover, the vulnerability of Southeast Asia's sea lanes, namely the Straits of Malacca, Sunda, and Lombok, and the passage into the South China Sea, give rise to concern as the oil-import dependence of China has already grown to 55 per cent, well above the critical level based on international standards.[26] The reality of China's growing energy import dependence has alerted Beijing to the importance of securing stable alternative suppliers of energy, since heavy reliance on Iran and other politically unstable countries for energy resources carries real risks and high cost.

China, therefore, needs to adjust its "going out" strategy and place a high priority on getting as much future oil and gas as possible from as close to home as possible so as to ease some of its dependence on insecure transit routes and unstable countries. In the context of China's attempts to diversify the sources of energy and raw materials, its immediate neighbours are seen as especially important.[27] Although the region does not help directly to increase the diversifying supplies away from the geopolitically insecure Middle East, the region is critical because it acts as the principle route of transportation of Chinese oil imports.[28] In this regard, Southeast Asia has become an important player.

Southeast Asia is also a crucially important area in which China's energy diversification strategy has been targeted. Despite the region's decline in oil exports, this producer-consumer relationship remains in place today because abundant LNG (liquefied natural gas) has given Southeast Asia new resource wealth. In geopolitical terms, Southeast Asia is home to some attractive resources and regulated markets, vulnerable transportation routes, and a new robust competition for regional resources. Bilaterally, ASEAN, as the fourth largest LNG producer in the world, provides China an important opportunity to shift its energy import structure and to move to a cleaner alternative energy option. China has financial capital, energy demand, and an interest in diversifying its energy supply, and can leverage its investment to increase the supply of resources on global markets. Finding opportunities for joint LNG projects is mutually beneficial to China, ASEAN states, and other Asian countries. This situation makes China's future relations with this gas-rich ASEAN a critical long-term and win-win goal.

China's "Less Coal, More Gas" Strategy

As analysed in Chapter 2, China's economic growth model is "state centric", leading to its heavy dependence on energy, especially on coal. According to the BP Energy Outlook for 2030, China accounted for 80 per cent of the growth in global coal demand between 1990 and 2010, and it is expected to account for 77 per cent to 2030.[29] Continued over-reliance on coal poses a great many problems and challenges for China. Although the country's coal reserves are immense, growing concerns over mining safety and, more importantly, the well-publicized local pollution and environment damages have pressed Chinese government to seek diversification of the energy mix, greater efficiency in the use of coal, and tighter environmental controls on coal-fired power generation.

China's Twelfth Five-Year Plan (2011–15) maps a path for the use of cleaner energy sources to mitigate the effects of rapidly rising energy demand. The plan also sets a target for reducing carbon intensity to 40–45 per cent below 2005 levels by 2020, in line with China's Copenhagen pledge. As China sees natural gas as a critical "bridge fuel" to help reduce its reliance on coal, the plan establishes that gas has to be aggressively promoted, targeting an 8.3 per cent share in the primary energy mix in 2015. According to the National Development and Reform Commission (NDRC), China plans to more than double its gas consumption from about 107.5 billion cubic meters (bcm) in 2010 to roughly 260 bcm in 2015.[30] To achieve this target, China needs to import 50–60 bcm of natural gas in 2015, similar to the current import level of the European Union.[31]

Although the Chinese government is optimistic about its domestic shale gas and has made ambitious development plans, there still exist many problems that make it difficult to achieve production targets. For example, many Chinese state and private companies lack the technological expertise to properly conduct horizontal drilling and require partnerships with foreign entities or acquisition of foreign technology. Additionally, although China technically holds shale reserves that are estimated to be more than 1,000 trillion cubic feet, it is not clear that all these reserves are economical to explore and develop.[32] CNOOC estimates that it costs around US$15 million to develop a single well. As such, China's National Energy Administration, the country's energy regulatory and planning agency, said that China has revised its 2020 forecasts for domestic shale gas production to 30 billion cubic meters per year, down from a previous target of 60 to 100 billion cubic meters. This revision acknowledges the complexity, and at times difficulty, involved in developing China's shale gas resources. China has to move to secure access to large volumes of natural gas from a number of different sources.

Beijing already depends on imports for 30 per cent of its gas needs and is likely to rise to 50 per cent import dependence by 2020. China has to find supply options and increase gas imports, either through pipelines or through LNG. The efforts to double gas consumption will require China to increase access to global gas supply through pipelines and LNG, since domestic production is unlikely to be sufficient.

China's Pipelines Network

By the end of 2010, the total length of domestic natural gas pipelines reached 40,000 km, including the east section of the second west-east gas pipeline, the Shanxi–Beijing gas pipeline, and the Sichuan-east gas pipeline, which basically formed a backbone pipeline network in China.[33] The west-east pipelines have several phases, with the first phase delivering gas to the Yangtze River Delta and the second to the Pearl River Delta in southern China. Both phases are now in operation. During the Twelfth Five-Year Plan period (2011–15), the Chinese oil and gas pipelines are still in rapid expansion with their systems becoming more complete, and it is expected that the total length of the long-distance oil and gas pipelines will reach over 100,000 km by the end of the Twelfth Five-Year Plan.[34]

Currently international gas is delivered to China through several multinational gas pipelines. The Turkmenistan–China pipeline ends in Xinjiang (also called the Central Asia–China pipeline) and later connects to the domestic west–east pipeline that sends gas from Xinjiang to the coastal provinces. This energy linkage to Turkmenistan via Uzbekistan and Kazakhstan has many implications as it forms a part of China's broader Central Asia Strategy which is to enhance the region's economic integration with Chinese markets through the pipeline construction.

The other major planned pipelines are from Myanmar and Russia. The US$2.5 billion China–Myanmar pipeline had begun operations in June 2013 and, once at full capacity, could deliver up to 12 bcm of gas every year to the southeastern Chinese provinces. This pipeline is designed to receive gas from the Middle East and Africa in order to bypass the Straits of Malacca which proponents in China argue can reduce shipping costs and enhance security. More importantly, it also has strategic implications of economic and energy cooperation with Myanmar and other ASEAN countries.

Since the early 1990s, China has been in discussion with Russia on several pipeline projects, with the aim of importing natural gas from Russian gas fields such as Kovyktinsoye in Irkutsk, West Siberia (through the planned Altai Pipeline), Sakha in East Siberia, and Sakhlin. In May 2014, the two countries signed a US$400 billion pact, under which Russian gas giant Gazprom will supply China with 38 billion cubic meters of gas annually for thirty years, starting in 2018, through the planned "eastern pipeline", which will connect Russia to China's north-eastern Heilongjiang province, and from there connect with the industrialized cities along Chinese eastern coast.[35] Under the pact, Gazprom will be responsible for upstream exploration, gas processing, and pipeline construction with Russia, and its Chinese partner, China National Petroleum Corporation (CNPC), will build pipelines inside China and all supporting and storage facilities.

In June 2012, CNPC signed a framework agreement with the government of Afghanistan for a feasibility study on a pipeline from Turkmenistan to China via Afghanistan.[36] CNPC views this proposed project as part of a larger pipeline that stretches from Iran, where CNPC has a contract to develop Phase II of South Pars, the world's largest natural gas field.[37] In 2011, the Iranians froze CNPC's US$4.7 billion contract due to its failure to start work. However, if CNPC eventually exploits South Pars, a pipeline from the field to China via Turkmenistan and Afghanistan would be an important alternative for CNPC to transfer its production back to China.

Such pipeline projects and investments are in conjunction with China's long-term goal of developing its volatile western region, an area of priority in China's Twelfth Five-Year Plan (2011–15). China's booming economic growth in the past decades has created lopsided economic development on the national level. Much of the western region has been left far behind and is vastly underserved by internal gas and oil distribution networks. These projects would accelerate the Grand Western Development, a national strategy launched in 2000 to promote the growth of China's western provinces. In the long run, Beijing hopes that these transnational pipeline projects may play into the larger plans to extensively increase its presence and connections in Central Asia and the Middle East, hence ensuring access to the region's natural resources as well as diversifying energy transportation channels.

China's LNG Imports

Although China favours overland pipelines for gas supplies because it feels pipelines are more secure, the impact of these pipeline projects on China's overall energy picture is still small, particularly given China's projected oil and gas imports. The Central Asia–China and Myanmar gas pipelines together will have a final transport capacity of 52 bcm, or roughly 40 per cent of China's projected import demand in 2030.[38]

Moreover, transnational pipelines require a considerable amount of financial capital, implying significant amount of risk in both the political and economic sense. Given the political transformation in Myanmar, Beijing may consider that the political risk is too high to expand pipeline investment in a country that is undergoing dramatic changes, especially when other options exist for importing gas. Moreover, the difficult talk with Russia likewise is leading Beijing to rethink the security and cost of linking to a pipeline from its northern neighbour. Therefore, China is to rely more on LNG imports because such imports are potentially cheaper and of fewer risks than transnational pipelines.

LNG is imported from abroad, then stored, re-gasified, compressed, and distributed to special customers through long gas pipelines, or sold through truck loading stations. As gas demand, driven by power demand and residential use, is likely to further increase in coastal China with the continued expansion of the city gas network, based on long-term contracts signed with overseas gas suppliers, it is less costly to import LNG than to rely on transporting domestic gas across the country.[39]

Chinese companies have already signed many LNG contracts, and have invested extensively in LNG terminals, with four currently in operation — along the coast in Fujian, Guangdong, Zhejiang, and Liaoning — and as many as eleven planned terminals could become operational by 2015.[40] According to CNPC, from 2003 to 2005, China's annual LNG imports were less than 500 tons. China began to increase LNG imports in 2006, and its LNG imports could reach 16.5 million tons in 2013. If the planned terminals all come into operation as expected, China should have an import capacity of 87 million tons, more than five times the current level of LNG imports.[41]

Chinese national oil companies are investing in Asia, Canada, Australia, and many other LNG projects around the world. Without a doubt, this gas development strategy will draw China closer to Australia, the Middle East, and Southeast Asian suppliers, creating potential opportunities for gas cooperation between China and some ASEAN countries, such as Indonesia and Malaysia.

ASEAN's LNG is tempting to China because it is closer to China's centres of gas demand than China's domestic resources or Central Asian resources are. Chinese government seems to have committed to a future of using ASEAN's LNG to service China's gas demand which are mainly concentrated in the coastal cities.

OPPORTUNITIES FOR SOUTHEAST ASIA

Bilateral energy cooperation between China and the ASEAN countries is not new. China–ASEAN energy cooperation was initiated in the late 1970s. In July 1978, China and the Philippines signed a long-term Petroleum Trade Agreement which was the first energy trade agreement between China and an ASEAN member country.[42] Since 2000, China has become Brunei's eighth largest consumer of crude oil, and in 2003 and

2006, China signed major contracts with Indonesia and Malaysia for the supply of LNG.[43] China sees Southeast Asia as contributing to its energy resource security by offering coal and natural gas exports. Chinese policymakers have been thinking of building pipelines from Myanmar and Thailand to end up in Yunnan, meaning that oil tankers need to only unload in Myanmar and Thailand ports without having to traverse regional straits. While ASEAN countries see China as contributing to their energy resource development and climate security by offering capital and clean energy technologies, there is great potential for energy cooperation between China and Southeast Asia.

ASEAN's Shifting Energy Resources

Southeast Asia has large reserves of oil and natural gas (see Table 3.2), and has long played an important role as an exporter of oil and gas. It is home to many attractive resources, vulnerable transportation routes, and a new robust competition for regional resources. Simultaneously, however, there has been a rise of energy nationalism among states, which has led to state-led competition for energy resources.

In history, Southeast Asia was under the exclusive control of Western powers during the colonial period. The oil-producing territories of Indonesia, Brunei, Sarawak, and Burma (Myanmar) were controlled by the Dutch or British. By the early years of the twentieth century, crude oil was being exported from what was then British Burma and Sarawak. Western control of and the head start of Western oil companies in the exploration, development, transportation, and refining of the petrochemical as well as natural gas industries had helped to entrench Western interests in this region. After World War II and especially in the 1970s, Japan managed to secure its access to the natural resources in Southeast Asia through its ODA (official development aid) programmes and investment activities, and Southeast Asia became one of Japan's important energy resource providers.

But this historical pattern has been changing as the region's energy demand grows strongly and the markets become more dynamic and diversified. Although ASEAN's oil production has been falling steadily and is projected to fall further, ASEAN as a whole is to witness a significant expansion in its gas production, drawing on a large resource base and growing demand for LNG in the Asia-Pacific market. In other words, as oil production declines, natural gas production is on the rise.

TABLE 3.2
Proven Reserves of Oil and Gas in Selected ASEAN Countries, at end 2012

	Oil			Natural Gas		
	Proven Reserves (billion barrels)	**Share of Total**	**R/P Ratio**	**Proven Reserves (trillion cubic feet)**	**Share of Total**	**R/P Ratio**
Brunei	1.1	0.1	19	10.2	0.2	22.9
Indonesia	3.7	0.2	11.1	103.3	1.6	41.2
Malaysia	3.7	0.2	15.6	46.8	0.7	20.3
Myanmar	0.05	*	N.A.	7.8	0.1	17.4
Thailand	0.4	*	2.7	10.1	0.2	6.9
Vietnam	4.4	0.3	34.5	21.8	0.3	65.6
Total World	1,668.9	100	52.9	6,614.1	100	55.7

Notes: N.A. — Not Available;
* — Less than 0.05%;
R/P (reserves to production) ratio — defined as the length of time, in years, that the remaining reserves would last at the prevailing annual rate of production.

Source: *BP Statistics Review of World Energy*, June 2013, pp. 6–20.

Southeast Asia is richer in natural gas than in oil. According to *BP Statistics Review of World Energy*, by the end of 2012, Indonesia's proven reserves of gas stood at 103 trillion cubic feet, accounting for 1.6 per cent of the world's total endowment; Malaysia's proven reserves of gas stood at 47 trillion cubic feet, accounting for 0.7 per cent of the world's total endowment. The region's total proven reserves of natural gas accounted for about 3.7 per cent of the world's total endowment. At current levels of production, ASEAN's proven reserves of natural gas would sustain production for another thirty-three years. Production of natural gas in Southeast Asia will increase from 203 bcm in 2011 to about 260 bcm in 2035.[44] Indonesia, the largest gas producer, will increase its gas output from 81 bcm to 140 bcm. However, the gas development in Southeast Asia also has some impediments, as major gas producers, such as Indonesia and Malaysia, face difficulties in allocating supply between domestic demand and exports.[45]

Apart from gas, ASEAN's coal output continues to expand, led by Indonesia, to meet fast-increasing domestic demand and export growth. According to IEA, at the end of 2011, ASEAN had 28 billion tons in total coal reserves, or 2.7 per cent of the world total.[46] ASEAN's coal production will increase from 348 Mtce (million tons of coal equivalent) to around 620 Mtce in 2035.[47] But coal production in ASEAN also faces challenges. As in other sectors, a key challenge is to ensure adequate investment to increase the coal production and related infrastructures to meet growing domestic demand and export expansion.

ASEAN's Dynamic Energy Trade

Compared with some of its neighbours, Southeast Asia is richly endowed with fossil and renewable energy resources though they are distributed unevenly across the region. Currently, ASEAN is an exporter in net energy-equivalent terms, as exports of coal and natural gas more than offset net imports of oil.

ASEAN as a whole produces about 40 per cent of the world's supply of LNG. In 2013, it had almost 90 bcm per year of LNG liquefaction capacity, accounting for almost one-quarter of the world total. This is located in Indonesia, Malaysia, and Brunei. Net exports of gas from the region are expected to increase in the medium term, approaching 68 bcm by 2020 (see Table 3.3). Indonesia and Malaysia remain net gas exporters, though Malaysia's net exports are significantly eroded by domestic demand growth.

TABLE 3.3
Fossil Fuel Net Trade by ASEAN Countries

	Oil (mb/d)			Gas (bcm)			Coal (Mtce)		
	2012	**2020**	**2035**	**2012**	**2020**	**2035**	**2012**	**2020**	**2035**
Indonesia	–0.6	–1.0	–1.4	42	56	58	251	363	385
Malaysia	0.1	–0.1	–0.4	22	30	17	–22	–33	–54
Philippines	–0.2	–0.3	–0.6	0	–1	–7	–12	–22	–40
Thailand	–0.6	–0.9	–1.5	–11	–30	–57	–26	–40	–67
Rest of ASEAN	–0.5	–0.8	–1.1	9	13	4	29	20	–6
Total ASEAN	–1.9	–3.1	–5.1	62	68	14	220	288	217

Source: IEA, *Southeast Asia Energy Outlook 2013*, p. 66.

Indonesia and Malaysia have the region's largest reserves of natural gas and presently account for about 70 per cent of Asia's gas trade. In 2012, Indonesia exported 42 bcm of gas, and this amount will increase to 56 bcm by 2035. Some 90 per cent of Indonesia gas is exported in the form of LNG, with 70 per cent going to Japan, 20 per cent to South Korea, and 10 per cent to Taiwan.[48] Malaysia's production of natural gas has steadily increased in recent years. In 2012, Malaysia exported 22 bcm of gas, and the figure will increase to 30 by 2020. Most of Malaysia's gas export goes to Japan, South Korea, and Taiwan. Brunei is the fourth largest producer of LNG in the world, with about 85 per cent of it is sent to Japan and 11 per cent to South Korea. Moreover, Indonesia is by far the dominant producer, having greatly increased its coal output and exports in the last decade. Since 2000, China has become Brunei's eighth largest consumer of crude oil, and in 2011, China's import of crude oil from Brunei was 612,000 tons (see Table 3.4).

On the supply side in Southeast Asia, there has been a lag in the growth of installed productive capacity in upstream oil and gas sectors to accommodate rising demand. ASEAN needs significant investment to related infrastructures so as to bring forward the projected amount of oil, gas, and coal production. According to the IEA, the region requires a total of US$705 billion of investment in fossil fuel-upply infrastructure (including pipelines, LNG terminals, and gas transmission and distribution infrastructure).[49] Financing investment in energy development and energy-supply infrastructures has always been a big challenge for ASEAN countries. It is therefore expected that private domestic and foreign companies will play a significant role in developing the region's energy sectors and related infrastructures.

CHINA'S RESOURCE DIPLOMACY IN SOUTHEAST ASIA

It is not easy to determine to what extent China's diplomacy is shaped by its global search for oil and raw materials in order to fuel its booming economy. But given the extent that uninterrupted energy supply provides a foundation for continued economic growth, which helps to secure the legitimacy of the Chinese Communist Party (CCP), the importance of oil and gas diplomacy cannot be underestimated,[50] especially in its neighbouring countries. The increasing ODA (official development aid), low-interest loans for energy-related infrastructure building, the numerous visits made in recent years by top Chinese leaders to neighbouring countries and other parts of the world in search for raw materials and markets underscore this importance.

TABLE 3.4
ASEAN's Crude Oil Export to China
(10,000 tons)

	2000	2001	2002	2003	2004	2005	2006	2007	2008	2009	2010	2011	2012
Vietnam	315.9	336.2	354.3	350.6	534.8	319.6	87.3	49.6	84.2	102.5	68.3	85.4	74.5
Indonesia	464.1	264.5	323.8	333.4	342.9	408.6	212.2	228.3	139.2	323.4	139.2	71.7	170.9
Malaysia	74.4	90.0	164.9	203.1	169.2	34.8	11.3	49.8	89.3	223.0	207.9	171.5	111.4
Thailand	28.5	22.7	74.0	161.0	91.5	119.2	114.8	110.2	76.5	60.7	23.1	33.4	30.3
Brunei	27.6	75.4	129.6	135.9	88.3	50.2	41.8	40.3	7.9	52.6	102.4	61.2	40.5

Source: *China's Customs Statistics Yearbooks*, 2000–12.

China often calls its Asian neighbours "periphery countries" where China has for centuries engaged in competition for leadership and now joins in a new competition to access energy supplies close to home that could reduce its dependence on cross — ocean oil shipping lanes. To create a favourable international/regional environment for economic modernization after the late 1970s, Chinese leaders made a deliberate effort to devise an integrated regional policy, known as "zhoubian zhengce" (periphery policy) or "mulin zhengce" (good neighbour policy). Beijing's periphery policy was aimed at exploring the common ground with Asian countries in both economic and security areas.

The energy issue did not become a factor in China's periphery policy until the mid-1990s when growing energy deficits began interfering in China's efforts to sustain economic growth. This new variable has helped China develop cooperative relations with some of its neighbours, notably some countries in Central and Southeast Asia, through ODA or resource diplomacy. The strategically important oil and gas pipelines linking Kyaukpyu in Myanmar to Southwest China, which had been completed in May 2013, is a good example of the new foreign policy's effectiveness.

China's ODA programme dates back to 1950s, but its early ODA was mainly concentrated in Africa and Latin America in support of new nations "fighting against colonialism and hegemony".[51] In the mid-1990s, China instituted a series of reforms in its aid programme, explicitly linking foreign aid with economic cooperation and trade, as the economic situation in China had changed greatly. By the mid-1990s, China faced a situation similar to that of Japan in the 1970s: a booming industrial economy that was flush with cash due to expanding exports, but increasingly dependent upon imported oil and other strategic resources. China became a net oil importer in 1996. China overtook Japan as the second largest oil consumer next to the United States in 2003, and became the third largest oil importer after the United States and Japan in 2004.[52] In 2010, China surpassed the United States to become the largest energy consumer in the world, accounting for 22 per cent of global energy consumption.[53] In the coming decades China will not only consolidate its position as the world's largest consumer of energy but also remain the largest oil importer.Like Japan in the 1970s, China soon began to leverage its burgeoning cash due to expanding exports to secure excess to strategic energy resources.

China has not been alone in stepping up its resource diplomacy efforts. Japan, for example, has a long-standing track record of resource diplomacy, having used the strategy during the resource boom of the 1970s.[54] The Japanese government offered financial and diplomatic assistance to its industrial corporations to sponsor the development of new mining firms in Southeast Asia, Latin America, and Oceania. These mining firms were connected to the Japanese economy through long-term contracts and investment aids, which guaranteed Japan's resource security by ensuring their output preferentially supplied to the Japanese market.[55] Similar to Japan's earlier aid programme, China's subsidized loans are often part of a larger investment package aimed at securing access to key strategic resources in developing countries. For example, China provided its first subsidized loan to Sudan in 1996 to finance oil exploration through a joint venture with CNOOC. From 2000 to 2005, China's outward FDI to Africa and Latin America rose by 37 per cent and 45 per cent respectively, with a large increase in mining, particularly the oil sector.[56]

China launched its ASEAN resource diplomacy much later than Japan as it did not normalize its political relations with ASEAN countries until the early 1990s. It was a milestone when ASEAN's perceptual change from "China as a threat" to "China as an opportunity" took place during the Asian financial crisis in 1997. In the crisis, China had firmly committed not to devaluate its currency and provided timely and generous financial assistance to ASEAN countries. China actively adjusted its ASEAN policies. Among major countries in the region, China was the first to sign the Treaty of Amity and Cooperation with ASEAN; the first to propose and sign an agreement to establish a FTA with ASEAN; the first to forge a Strategic Partnership for Peace and Prosperity with ASEAN; the first to pledge the accession to the protocol to the Treaty on Southeast Asia Nuclear Weapon Free Zone; and the first to sign a "strategic partnership" with ASEAN at the October 2003 Bali Summit.

In 2008, China announced the establishment of the China–ASEAN Investment Cooperation Fund and committed loans for ASEAN countries amounting to US$15 billion.[57] In 2013, China proposed to establish the Asian Infrastructure Investment Bank (AIIB),[58] presumably by drawing on its foreign reserves. China has also utilized the framework of the Greater Mekong Subregion Economic Cooperation Program, facilitated by the Asian Development Bank, to improve the physical connectivity between the Mekong region and Yunnan Province and Guangxi Zhuang Autonomous Region. China's diplomacy toward ASEAN turned out to be the most successful story of Chinese foreign policy in the post-Cold War era.[59]

It was under such a background that China was able to catch up with or even surpass Japan in its economic relations with ASEAN, and its aid to ASEAN countries increased continuously, with many going to resource-related infrastructure sectors. According to the China's 2014 white paper on foreign aid, from 2010 to 2012, China provided US$14.4 billion for foreign assistance in three types: grants, interest-free loans, and concessional loans, with Africa and Asia as the main beneficiaries and accounting for 52 per cent and 31 per cent of China's foreign assistance respectively, while economic and energy resource-related infrastructures accounting for 45 per cent of China's total foreign aid.[60] Although the white paper does not provide data by country, it is obvious that the amount of trade, investment, and ODA directed to Southeast Asia — in particular via infrastructure financing — has grown substantially in recent years, and that China is one of the largest sources of economic assistance, defined broadly, in Southeast Asia.[61]

China has also cultivated economic relations with states neighbouring the South China Sea. In Vietnam, China has helped develop railway construction, hydropower development and ship-building facilities. In the Philippines, China has invested in infrastructure, energy, agriculture, and mining. China's ODA to the Philippines grew from US$35 million in 2001 to US$1.14 billion in 2010, reflecting its close ties with the Arroyo administration.[62] China and the least developed mainland ASEAN countries are becoming increasingly economically integrated, although each of these countries has also sought to hedge against China's rising influence. China is the primary supplier of economic assistance to Myanmar, Cambodia, and Laos, financing a number of energy-related, infrastructure, agricultural, and other development projects in these countries.

While acknowledging that China is engaging in resource diplomacy in Southeast Asia, China insists that its external resource cooperation has been conducted under the win-win principle, as a Chinese official said, "China is enabling resource-endowed countries to obtain wealth, transferring resource exploration and development technologies to these countries, enhancing their employment and revenues by processing resource-based products in these countries, and helping achieve trade bilateral trade balances".[63] In China's consideration, many of China's aid programmes (such as investment in infrastructures) in Southeast Asia are to provide relatively greater and diversified long-term diplomatic benefits and comparatively fewer short-term economic ones.

CHINA'S FDI AND ENERGY COOPERATION

China's investment in oil and gas exploration and energy-related infrastructures is another important aspect in China–ASEAN energy cooperation. China's "resource diplomacy" and ASEAN's preferential policies for foreign investment after the financial crisis have stimulated Chinese state-owned and, increasingly, private-owned companies to invest in resource-related infrastructures and gas and coal exploration sectors.

Expansion of China's OFDI in Southeast Asia

ASEAN Becoming More Attractive to FDI

As discussed in Chapter 1, due to increasing regional integration, Southeast Asia is becoming a more attractive destination for FDI. As the ASEAN Economic Community is to be realized by the end of 2015, ASEAN has made important strides in the area of investment cooperation, in the form of ASEAN "one-stop investment centres" and the ASEAN Investment Area (AIA) which seeks to ensure national treatment for ASEAN investors by 2010 and for all investors by 2020.[64] Since the implementation of AIA in 1998, ASEAN has been committed to strengthening its investment regime, and this commitment was further reinforced with the signing of ASEAN Comprehensive Investment Agreement (ACIA) in 2009.

As most multi-national companies (MNCs) in ASEAN are from non-member countries and ASEAN countries are actively courting FDI, legislation is emerging or is in place in a number of member countries that would accord national treatment for all investors. Singapore, for example, does not discriminate between ASEAN and non-ASEAN investors. This largely leads MNCs to consider ASEAN as an integrated network in which transaction costs are low and with potential for an efficient regional division of labour, could render ASEAN far more attractive to foreign investors. This has been verified by the fact that the total FDI inflows to ASEAN reached a record of US$116.6 billion in 2011 from US$92.8 billion a year ago (see Figure 3.2).

Moreover, China–ASEAN FTA was established on 1 January 2010 with the signing of Merchandise Trade Agreement, Service Trade Agreement and Investment Agreement. China also offered an "early harvest programme" in 2004 which favoured Cambodia, Laos, Myanmar, and Vietnam (CLMV) and have greatly promoted trade in agriculture and marine products. Under the investment agreement reached in August 2008, China and ASEAN will establish a free, convenient, transparent and competitive investment regime. Obviously, the institutionalized China–ASEAN integration process has not only fuelled the China–ASEAN bilateral trade, but also led to the expansion of China's FDI to ASEAN.

FIGURE 3.2
World FDI Inflows to ASEAN and China
(US$ million)

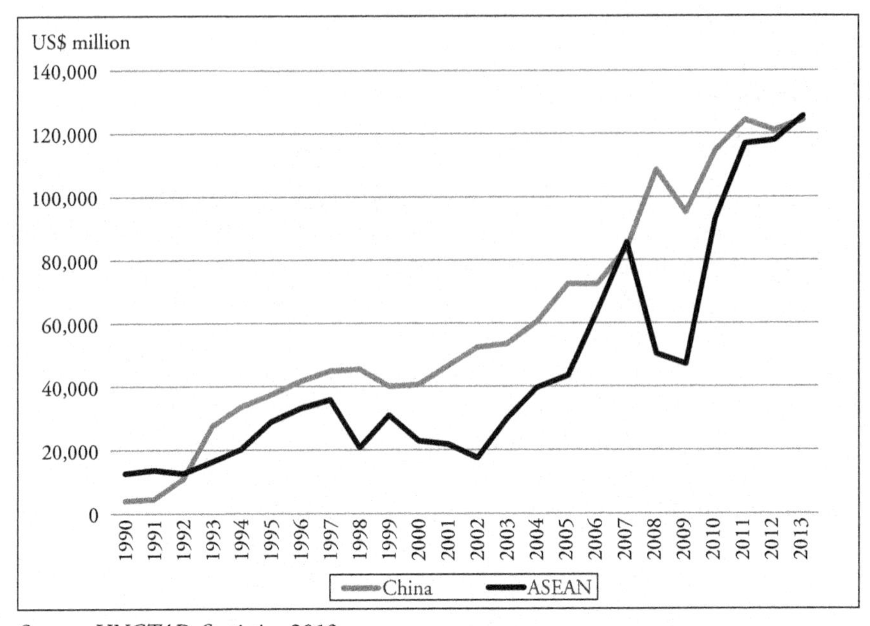

Source: *UNCTAD Statistics 2013*.

China's Investment in Southeast Asia

In the early 1990s, when China started expanding its outward foreign direct investments (OFDI), ASEAN member countries were receiving around 3 per cent of China's total OFDI. As developing countries' share rose over time, ASEAN's also rose dramatically. According to statistics from the Ministry of Commerce of China, between 2003 and 2013, ASEAN saw an increase of China's inward investment from US$0.12 billion to US$7.3 billion, with total accumulation being US$35.7 billion by the end of 2012, and ASEAN's share in China's total OFDI rose from 4.1 per cent to 6.7 per cent. Compared with the previous year, in 2011, China's OFDI in ASEAN increased 34.1 per cent, while its OFDI in the European Union (US$7.6 billion) increased 26.8 per cent, in the United States (US$1.8 billion) increased 38.5 per cent, in Russia (US$7.2 billion) 26.1 per cent. Compared with other big countries, China's share in ASEAN's total FDI inflows had increased from 1.9 per cent in 2004 to 5.3 per cent in 2011, being lower than that of Japan (13.2 per cent), but higher than that of the United States (5.1 per cent) and Australia (1.2 per cent).[65]

In the ASEAN-10 countries, some resource-rich ASEAN members witnessed a rapid expansion of China's OFDI inflows. For example, from 2003–12, the amount of China's OFDI in Myanmar increased from US$4.1 million (in 2004) to US$749 million; in Indonesia it increased from US$26 million to US$592 million; in Laos it increased from US$0.8 million to US$809 million; in Vietnam increased from US$12.8 million to US$349 million (see Table 3.5).

Although Chinese OFDI in Southeast Asia shows that the importance of seeking market opportunities and labour-division is greater than the importance of obtaining natural resources, resources-seeking is still one of the main motivations of Chinese OFDI. The figures published by Chinese Ministry of Commerce in 2013 show that of China's total OFDI in Southeast Asia in 2012, the share of mining (including oil and natural gas, ferrous metal and non-ferrous metal mining) had increased from 9.7 per cent in 2004 to 28.1 per cent in 2012 (see Figure 3.3), which was almost equal to that in China's total investment in Africa. In 2011 the mining sectors accounted for 30.6 per cent of China's total OFDI (US$2.52 billion)

FIGURE 3.3
Distribution of China's FDI in Southeast Asia, 2012

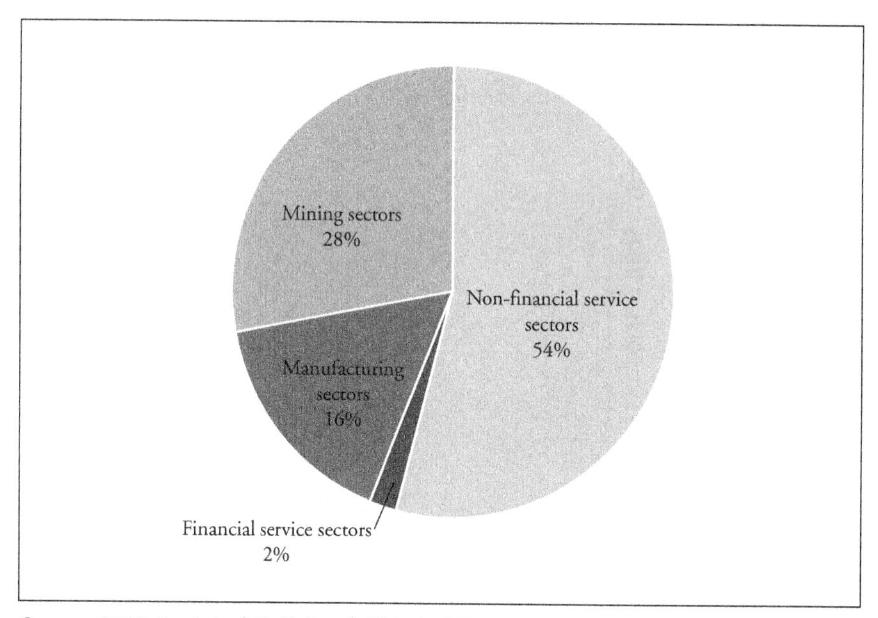

Source: *2012 Statistical Bulletin of China's OFDI*, p. 23.

TABLE 3.5
Chinese OFDI in ASEAN
(US$ million)

	2003	2004	2005	2006	2007	2008	2009	2010	2011	2012	2013	Accumulation by the Year 2013
Brunei	—	—	1.5	—	1.2	1.8	5.8	16.5	20.1	0.99	8.5	72.1
Cambodia	22.0	29.5	5.2	9.8	64.5	204.6	215.8	466.5	566.0	559.7	499.3	2,848.6
Indonesia	26.8	62.0	11.8	56.9	99.1	174.0	226.1	201.3	592.2	1361.3	1,563.4	4,656.7
Laos	0.8	3.6	20.6	48.0	154.4	87.0	203.2	313.6	458.5	808.8	781.5	2,770.9
Malaysia	2.0	8.1	56.7	7.5	–32.8	34.4	53.8	163.5	95.1	199.1	616.4	1,668.2
Myanmar	—	4.1	11.5	12.6	92.3	232.5	376.7	875.6	217.8	749.0	475.3	3,569.7
Philippines	0.95	0.05	4.5	9.3	4.5	33.7	40.2	244.1	267.2	74.9	54.4	692.4
Singapore	–3.2	48.0	20.3	132.2	397.7	1,551.0	1,414.3	1,118.5	3,269.0	1,518.8	2,032.7	14,750.7
Thailand	57.3	23.4	4.8	15.8	76.4	45.5	49.8	699.9	230.1	478.6	755.2	2,472.4
Vietnam	12.8	16.9	20.8	43.5	110.9	119.8	112.4	305.1	189.2	349.4	480.5	2,166.7
ASEAN	119.4	195.7	157.7	335.6	968.2	2,484.3	2,698.1	4,404.6	5,905.2	6,100.55	7,267.2	35,668.4

Source: Ministry of Commerce of China, *2013 Statistical Bulletin of China's Outward FDI*.

in Africa.[66] Considering the total amount of China's OFDI in ASEAN is larger than that in Africa, we can assume that the mining sectors are becoming increasingly attractive to China's investment.

To a large degree, the fast-growing Chinese OFDI in Southeast Asia also reflects the trends — China's growing demand for energy (primary coal and LNG) and raw-materials. Resource-rich countries like Myanmar, Indonesia, and Malaysia have attracted China's investment in the primary sector. In Myanmar, China's three largest state-owned oil companies — CNPC, Sinopec and CNOOC — have all started oil exploration projects there. In November 2008, CNPC and the Ministry of Energy Myanmar signed an agreement to build a US$2.3 billion crude oil pipeline and US$2 billion natural gas pipeline. After the Myanmar–China gas pipeline became operational in 2013, Myanmar will supply 10 billion cubic metres of gas to China per year, accounting for around 6 per cent of China's total gas consumption in 2013 (167.6 billion bcm).[67]

Impacts on Regional Energy Cooperation

From a regional perspective, China's FDI in Southeast Asia, especially in its energy sectors, will undoubtedly enhance regional energy cooperation between China and ASEAN countries. Chinese companies can play a key role in the transformation of ASEAN's power infrastructure.

As mentioned above, Southeast Asia has large reserves of oil and gas, and has long played an important role as an exporter of oil and gas. But this pattern has already changed as the region's oil demand grows ever more strongly. The region's oil output has been falling steadily, since peaking at 2.9 mb/d (million barrels per day) in 1996. Southeast Asia as a whole is currently a net oil importer, and individual countries' net-oil import dependence has been steadily rising. Southeast Asia's proven reserves of natural gas stood at 6.6 tcm (trillion cubic metres) at the start of 2009, or 3.7 per cent of the world's total endowment.[68] However, while the region remains an important supplier of LNG, gas is also increasingly sought to support power generation and industry in domestic markets. If no more new investment is increased and no new gas field is found, the surplus of supply over demand is expected to narrow and diminish by 2030.[69]

Thus it is greatly needed to enhance regional energy cooperation and introduce more foreign investment to energy sectors. An integral part of ASEAN's regional economic cooperation focuses on the energy sector where it is ASEAN's declared intention to ensure greater security and sustainability of regional energy supplies. In 2002, ASEAN member states adopted the ASEAN MoU on the Trans-ASEAN Gas Pipeline (TAGP). Once realized, this TAGP will have the potential of linking almost 80 per cent of the ASEAN region's total gas reserves, and will embody a far-reaching expression of the region's energy interdependence.[70] The TAGP is a massive project that would connect Southeast Asia in one of the largest networks of its type in the world. Linking the gas reserves of Indonesia, Malaysia, Singapore, Vietnam, Myanmar, the Philippines, Brunei, and Thailand, the project is backed by the major oil and gas companies in each of these countries and was originally projected to be in full operation by 2020.

The first cross-border gas pipeline in ASEAN region is the one exporting gas from Malaysia to Singapore and was commissioned in 1991. To date, eleven bilateral connections have been established and several more are in the process of design and construction. No doubt, China's expanding FDI in Southeast Asia can enhance full interconnection of these pipelines, and create an interconnected gas grid and increase gas trade throughout ASEAN-10 countries and beyond. For example, given the ambitious magnitude of TAGP and China–Myanmar pipelines, it is possible that the network may be extended into China and beyond, as some advocates of the TAGP argued in the 1990s that the network would eventually connect with gas markets in China, Japan, and India, making it the largest pipeline network in the world.

From a regional perspective, the pipeline project will undoubtedly enhance energy especially gas cooperation between China and Southeast Asia. Southeast Asia has large reserves of oil and natural gas, and has long played an important role as an exporter of oil and gas. But this historical pattern has changed as the region's demand grows ever more strongly. The region's oil output has been falling steadily, since peaking at around 2.9 million barrels per day (mb/d) in 1996. Oil production is projected to drop to 2.4 mb/d in 2015 and 1.4 mb/d in 2030. Southeast Asia as a whole is currently a net oil importer.

Southeast Asia's proven reserves of natural gas stood at 6.6 tcm (trillion cubic metres) at the start of 2009, or 3.7 per cent of the world's total endowment. As for individual countries, Indonesia's output of gas in 2010 was 82 bcm (billion cubic metres), consumption was 40.3 bcm; Malaysia's output was 66.5 bcm, consumption was 35.7 bcm; and Brunei's output was 12.2 bcm.[71] They are the main gas exporters among ASEAN countries. However, while the region remains an important supplier of

LNG, gas is also increasingly sought to support power generation and industry in domestic markets. Although gas production in Southeast Asia as a whole is projected to increase from 203 bcm in 2008 to 248 bcm in 2030, if no more new investment is increased and no new gas field is found, the surplus of supply over demand is expected to narrow from 63 bcm in 2008 to just 10 bcm by 2030.[72] Thus there is a great need to enhance regional energy cooperation and increase investment in energy sectors.

FROM COOPERATION TO CONFLICTS?

As China expanded its FDI to Southeast Asia (including energy resource sectors) after the global financial crisis in 2008, China–ASEAN energy cooperation developed to a new level, extending from energy trade to energy resource exploration and related infrastructure-building. However, while some momentum exists towards continued cooperation and energy cooperation has largely broadened China–ASEAN bilateral relations, some factors are pushing the region towards unhealthy competition or even conflicts, adversely affecting bilateral relations. Rising demand for energy resources has given rise to "energy protectionism" or "energy resource nationalism", leading China and Southeast Asia in a full-fledged competition to attract energy supplies. The geopolitical nature of China's relations to Southeast Asia relative to some of its neighbours (e.g., Japan, India) adds another source of possible tension as China seeks to establish a network of basing and logistical networks to strengthen its presence in the region. As such, China and ASEAN countries want energy cooperation to secure their own individual energy supplies and interests, rather than a sort of cooperation that leads to regional energy security.

Chinese policymakers, for example, have primarily viewed energy security as an ability to rapidly adjust their new dependence on global markets and to diversify the energy sources and routes. China sees Southeast Asia as contributing to its energy security by offering coal and gas exports, and sees Myanmar as an alternative port for oil and gas imports without having to traverse regional straits. In the eyes of Western observers, "buying stakes in foreign oil fields, militarily protecting vulnerable shipping lanes, and all-out 'energy scramble' for resources are key features of China's current approach to energy security".[73] In Southeast Asia, the situation is not much different as John Lee described that

> in the eagerness to deploy Chinese capital and expertise for rapid completion of resource extraction, transportation and power-generation projects, Chinese SOEs have been given wide leeway in disregarding environmental standards and the interests of local people affected by these projects.[74]

This is reflected in Myanmar's view which believes that China's halfhearted engagements at the local level and poor crisis management have added to the widespread perception that China is solely concerned about the security of its own business operations and energy security, while ignoring the needs and interests of ethnic nationalities.[75] For Vietnam, bauxite exploitation projects in Tay Nguyen (Central Highlands) to supply China have triggered significant civil society movements and become a major turning point in the domestic politics of Vietnam,[76] as it has activated a movement of solidarity among scientists, intellectuals, and civil society groups in Vietnam.[77]

As such, China's presence in Southeast Asia has created more negative impacts than positive ones on the local communities. It has not brought substantial benefits but instead strengthened the locals' perception that their economic relations were unequal and asymmetrical. In particular, locals believe that China is grabbing their natural resources and disregarding their interests.[78] Some observers are concerned that China will also replicate the sort of "neo-mercantilist" strategies that Japan adopted during its high-growth phase, a possibility that is reinforced by the prominence of "state capitalism" in a number of rising powers.[79]

Nevertheless, a concern with protecting national interests is hardly exclusive to China; on the contrary, it is an increasingly prominent feature of resource politics across the globe.[80] In Southeast Asia, some countries seem equally preoccupied on placing national interests and domestic energy security concerns well above bilateral and regional cooperation, leading to resource nationalism on the rise. Consequently, foreign investors in natural resources face risks as well as opportunities in the region. The following two chapters will further analyse this issue through the case study of Myanmar and Indonesia. These two countries are chosen as the case study of this book because Myanmar borders China and is in a transition period, its complex attitude to China's investment is representative of those mainland Southeast Asian countries; while Indonesia is an emerging big economy. Its energy relations with China should be looked at in the context of a broader bilateral political and economic relationship.

Notes

1. Flynt Leverett, "Resource Mercantilism and the Militarization of Resource Management", in *Energy Security and Global Politics*, edited by Daniel Moran and James A. Russell (Abingdon: Routledge, 2009), pp. 211–41.
2. Erica S. Downs, "China's Quest for Overseas Oil", *Far Eastern Economic Review*, September 2007.
3. Francis McGowan, "Can the European Union's Market Liberalism Ensure Energy Security in a Time of 'Economic Nationalism'?", *Journal of Contemporary European Research* 4, no. 2 (2008): 90–106.
4. Philip Andrews Speed, Xuanli Liao, and Roland Dannreuther, "The Strategic Implications of China's Energy Needs", *Adelphi Papers* 346 (London: International Institute for Strategic Studies, 2002).
5. Zhao Suisheng, "China's Global Search for Energy Security: Cooperation and Competition in the Asia-Pacific", *Journal of Contemporary China* 17, no. 55 (2008): 207–27.
6. Flynt Leverett, "Resource Mercantilism and the Militarization of Resource Management", in *Energy Security and Global Politics*, edited by Daniel Moran and James A. Russell (Abingdon: Routledge, 2009), pp. 211–41.
7. Jeffrey D. Wilson, "Northeast Asian Resource Security Strategies and International Resource Politics in Asia", *Asian Studies Review* 38, no. 1 (2014): 15–35.
8. Stuart Harris, "Global and Regional Orders and the Changing Geopolitics of Energy", *Australian Journal of International Affairs* 64, no. 2 (April 2010).
9. Flynt Leverett, "Resource Mercantilism and the Militarization of Resource Management", in *Energy Security and Global Politics*, edited by Daniel Moran and James A. Russell (Abingdon: Routledge, 2009).
10. John Lee, "China's Geostrategic Search for Oil", *The Washington Quarterly* 35, no. 3 (2012).
11. Heinrich Kreft, "China's Quest for Energy", *Policy Review*, no. 139 (1 October 2006).
12. Susan Shirk, *China: Fragile Superpower* (New York: Oxford University Press, 2007), p. 138.
13. Mark Beeson, Mills Soko, and Wang Yong, "The New Resource Politics: Can Australia and South Africa Accommodate China?", *International Affairs* 87, no. 6 (2011).
14. U.S. Energy Information Administration, "Country Briefs-China", <http://www.eia.gov/countries/cab.cfm?fips=CH> (accessed 22 March 2015).
15. Ibid.
16. Li Geqin, "Zhongguo nengyuan waijiao xin wenti jiqi duice" [New Problems of China's Energy Diplomacy and Counter Measures], *Contemporary World* 3 (2008).

17. Quoted from Zhao Hongtu, "The Myth of China's Overseas Energy Investment", *East Asian Forum*, 4 March 2015.
18. Ministry of Commerce of China, *2012 Statistical Bulletin of China's Outward FDI*, p. 41.
19. "China's CNPC Targets Overseas Integration Deals", *Petroleum Intelligence Weekly*, 23 January 2006.
20. *China Customs Statistics Yearbook 2011*.
21. Philip Andrews-Speed, "Do Overseas Investment by National Oil Companies Enhance Energy Security At Home? A View From Asia", in *Oil And Gas For Asia*, NBR Special Report #41, September 2012.
22. Chen Mo, "Zhongguo nengyuan anquan xin sikao" [New Thoughts on China's Energy Security], *Journal of West Asia and Africa*, no. 6 (2012).
23. *China Customs Statistics Yearbook 2012*.
24. In 2008, China's political stability index for crude oil import sources represented the level at 29, while that of Japan, Korea and the United States represented the levels at 44, 45, and 33 respectively. See APERC, "APEC Energy Review 2008", <http://www.ieej.or.jp/aperc/2008pdf/Overview2008.pdf> (accessed 20 April 2013).
25. Zha Daojiong, "China and Iran: Energy and/or Geopolitics", in *Oil And Gas for Asia*, NBR Special Report #41, September 2012.
26. "Woguo 2020 nian zhanlue shiyou chubei liang jiang da shijie dier" [China's Strategic Oil Reserves Will Reach the Second Largest in the World], <http://www.china5e.com/show.php?contentid=207204> (accessed 25 March 2015).
27. You Ji, "Dealing with Malacca Dilemma: China's Efforts to Protect Its Energy Supply", *Strategic Analysis* 31, no. 3 (May 2007).
28. Philip Andrews-Speed and Roland Dannreuther, *China, Oil and Global Politics* (Abingdon: Routledge, 2011).
29. "BP Energy Outlook for 2030", January 2013, <http://www.bp.com/content/dam/bp/pdf/statistical-review/BP_World_Energy_Outlook_booklet_2013.pdf> (accessed 25 March 2015).
30. "Tianranqi fazhan shier wu guihua" [Twelfth Five-Year Plan on Natural Gas],<http://zfxxgk.nea.gov.cn/auto86/201212/W020121203312244945303.pdf> (accessed 25 March 2015).
31. Commercial and Strategic Opportunities for LNG in China", <http://www.norway.cn/Global/SiteFolders/webbeij/DNV%20-%20China%20LNG%20Final%20Report.pdf> (accessed 15 November 2013).
32. EIA, "World Shale Gas and Shale Oil Resources Assessment", 17 May 2013, <http://www.eia.gov/analysis/studies/worldshalegas/pdf/chaptersi_iii.pdf> (accessed 25 March 2015).
33. CNPC Research Institute of Economics and Technology, "Report on Domestic and Overseas Oil & Gas Industry Development in 2010", January

2011, <http://www.ief.org/_resources/files/content/energy-outlooks/2010-cnpc-industry-report-20110124.pdf> (accessed 22 December 2013).
34. Ibid.
35. Du Juan, "Gas Deal Supplies Energy Diversification", *China Daily*, 22 May 2014.
36. Islamic Republic of Afghanistan, Ministry of Mines, "Framework for Construction of Turkmenistan Gas Pipeline via Afghanistan to Tajikistan and China Signed", 16 June 2012, <http://mom.gov.af/en/news/10587> (accessed 25 March 2015).
37. "Iran freezes CNPC's South Pars Phase II Contract", *Energy Intelligence*, 28 June 2012, <http://www.energyintel.com/Pages/ArticleSummary/771520/Iran-Freezes-CNPC-s-South-Pars-Phase-11-Contract> (accessed 25 March 2015).
38. John Seaman, "Energy Security, Transnational Pipelines and China's Role in Asia", *Asie.Visions 27* (April 2010).
39. In fact, transport bottlenecks have been a major reason that China has been a net coal importer for the last several years. Coal imported into Guangdong from Indonesia has been cheaper than coal from Shanxi, for example.
40. *CNPC Report 2011.*
41. "Commercial and Strategic Opportunities for LNG in China", DNV Clean Technology Center, October 2011, <http://www.taylor-dejongh.com/wp-content/uploads/2012/07/Natural-gas-in-China.pdf> (accessed 23 March 2013).
42. Li Tao and Liu Zhi, "Shixi Zhongguo dongmeng nengyuan hezuo" [An Analysis to China–ASEAN Energy Cooperation], *Journal of Southeast Asian Studies*, no. 3 (2006).
43. Ian J. Storey, "China a Major Player in South East Asia Pipeline Politics", *Straits Times*, 23 October 2009.
44. IEA, *Southeast Asia Energy Outlook 2013*, p. 65.
45. Ibid., p. 65.
46. Ibid., p. 69.
47. Ibid., p. 69.
48. *APEC Energy Overview 2013.*
49. IEA, *Southeast Asia Energy Outlook 2013*, p. 66.
50. Gerald Chan, "China Joins Global Governance: The 10 Conundrums", in *China and the New International Order*, edited by Wang Gungwu and Zheng Yongnian (Abingdon: Routledge, 2008), p. 176.
51. Shang Changfeng, "Wenge shiqi zhongguode duiwai yuanzhu" [Chinese Foreign Aid in the Era of the Cultural Revolution], *Dangshi wenhui* [Party History Digest] 2 (2010): 57.
52. David Zweig and Bi Jianhai, "China's Global Hunt for Energy", *Foreign Affairs* 84, no. 5 (2005).
53. *BP Statistical Review of World Energy*, June 2013.

54. Jeffrey D. Wilson, "Northeast Asian Resource Security Strategies and International Resource Politics in Asia", *Asian Studies Review* 38, no. 1 (2014): 15–35.
55. Ozawa Terutomo, *Multinationalism, Japanese Style: The Political Economy of Outward Dependency* (Princeton: Princeton University Press, 1979).
56. Takaaki Kobayashi, "Evolution of China's Aid Policy", Japan Bank for International Cooperation, Working Paper No. 27 (2008).
57. The China–ASEAN Investment Fund (CAF) is a private equity fund that commenced operations in 2010 to promote investment in ASEAN countries, with the China Exim-bank playing a central role. The International Finance Corporation, a member of the World Bank Group, became an investor in CAF in 2011 with the aim of improving the standards of Chinese funds.
58. On 24 October 2014, twenty countries signed a non-binding multilateral Memorandum of Understanding in Beijing, China to set up the AIIB, which will be funding infrastructure development. The proposed initial authorized capital for the AIIB is US$50 billion, of which 20 per cent will be the paid-up capital. The AIIB will use commercial funding from financial markets to raise the necessary resources to provide competitive loans to the member countries. AIIB's objectives would be to speed up infrastructure development in energy and power, transportation, telecommunication, rural infrastructure and agricultural development, urban development, logistics and any other productive sector as deemed appropriate. See *China Daily*, 24 October 2014.
59. Wang Jianwei, "Great Power Relations and Their Impact on Japan–Southeast Asian Relations", in *Japan's Relations with Southeast Asia: The Fukuda Doctrine and Beyond*, edited by Lam Peng Er (Abingdon: Routledge, 2013).
60. China's Official Development White Paper 2014, <http://news.xinhuanet.com/english/china/2014-07/10/c_133474011.htm> (accessed 26 July 2014).
61. Thomas Lum, Hannah Fischer, Julissa Gomez-Granger, and Anne Leland, "China's Foreign Aid Activities in Africa, Latin America, and Southeast Asia", *Congressional Research Service*, 25 February 2009, <http://www.fas.org/sgp/crs/row/R40361.pdf> (accessed 21 September 2013).
62. Aileen San Pablo-Baviera, "The Philippines in China's Soft Power Strategy", *ISEAS Perspective*, 13 June 2013.
63. Xu Yongfa, president of CNPC Economics and Technology Research Institute, as quoted in *Interfax, China Energy Weekly*, 27 February–2 March 2012, p. 8.
64. Michael G. Plummer, *ASEAN Economic Integration: Trade, Foreign Direct Investment, and Finance* (Singapore: World Scientific, 2008), p. 99.
65. ASEAN Secretariat, *ASEAN Statistical Yearbook 2012*, p. 131.
66. State Council Information Office, "China–Africa Economic and Trade White Paper", August 2013, <http://news.xinhuanet.com/english/china/2013-08/29/c_132673093_3.htm> (accessed 25 March 2015).

67. "Woguo chengwei disan da tianranqi xiaofei guo" [China became the third largest gas consuming country], *Zhongguo shiyou bao* [*China Petroleum Daily*], 20 May 2014.
68. *BP Statistical Review of World Energy*, June 2011.
69. IEA, *World Energy Outlook 2009*, p. 552.
70. Ibid., p. 566.
71. *BP Statistical Review of World Energy*, June 2011.
72. IEA, *World Energy Outlook 2009*, p. 552.
73. Benjamin K. Sovacool and Vu Minh Khuong, "Energy Security and Competition in Asia: Challenges and Prospects for China and Southeast Asia", in *ASEAN Industries and the Challenge from China*, edited by Darryl S.L. Jarvis and Anthony Welch (Basingstoke: Palgrave Macmillan, 2011).
74. John Lee, "Myanmar Pivots Awkwardly Away from China", *ISEAS Perspective*, 12 December 2013.
75. Bernt Berger, "China's Troubled Myanmar Policy", *The Diplomat*, 23 August 2013.
76. Huong Le Thu, "Vietnam: Straddling Southeast Asia's Device", *Trends in Southeast Asia* #5 (2014).
77. Hunter Martson, "Bauxite Mining in Vietnam's Central Highlands: An Arena for Expanding Civil Society?", *Contemporary Southeast Asia* 34, no. 2 (2012): 173–96.
78. Ibid.
79. Mark Beeson, Mills Soko, and Wang Yong, "The New Resource Politics: Can Australia and South Africa Accommodate China?", *International Affairs* 87, no. 6 (2011).
80. M. Stevens, "Rising Resource Nationalism A Worry of Miners", *The Australian*, 7 June 2011.

4

CASE STUDY (1)
Myanmar

Although Myanmar is a natural resource-rich country, from a global perspective, it does not have particularly notable hydrocarbon reserves. For China, however, from a regional perspective, Myanmar's gas reserves and strategic position are significant in terms of energy security and regional cooperation. However, although China has planned to stake a long-term strategic energy investment in Myanmar and its capital has largely poured into resources and energy-related infrastructural sectors, some factors are pushing the two countries' energy ties toward competition and conflicts. Chinese foreign direct investment (FDI) in Myanmar no doubt ties the two countries closer together, but also reminds politicians of "new colonialism", creating concerns and resource nationalism in the local society. Moreover, as Myanmar opens up, the entry or resurgence of more global players like Japan and India provides Myanmar with more options, facilitating its attempts at loosing itself from the Chinese grip. Hence, the questions that require examination are as follows: What direction is the China–Myanmar energy tie going in, cooperation or conflict? How will China adjust its approaches and better embrace the new environment and dynamic changes in the country?

GLOBAL ENERGY RESOURCE NATIONALISM

Over the past two decades, there has been a growing recognition that natural resources play an important and complex role in international and regional peace and security. Natural resources can be a source of cooperation and stability, it can as well be a source of hostility or grievances, which may be related to the loss of sovereignty, the inequitable distribution of benefits from natural resources, the lack of opportunities for local people, or environmental and social harm caused by unsustainable extraction of resources. While such hostility or grievances may not be the sole causes of conflicts, they may lead to the rise of resource nationalism which has real potential to spark armed conflicts especially in disputed areas.

The term "resource nationalism" is often defined as a national government's assertion of ownership rights over oil and natural resource reserves within their territorial boundaries, usually in ways that contradict liberal paradigms for encouraging foreign investment, and work against the international energy companies.[1] When sovereignty states seek control over natural resources for strategic, political, and economic reasons, it is referred to as resource nationalism. In discussing resource nationalism in the oil sector in the Middle East, Stevens defines the term as consisting of "two components — limiting the operations of private international oil companies and asserting a greater national control over natural resource development".[2] The drivers of resource nationalism can arise from either national circumstances (increasing domestic demand for natural resources) or international ones (increasing commodity prices).[3]

Other scholars believe that resource nationalism is a concept in which the natural resources in the ground or under the sea are the property of the nation rather than of a firm or individual who owns the surface area. In this view, natural resources are a "national patrimony" and, consequently, should be used for the benefit of the nation rather than for private gain.[4] In Asia, it is the rising demand for energy that is leading to growing energy resource nationalism and increasing heavy state-support control over energy resources.

On the supply side, the goal of resource nationalism is to control the natural resources and maximize rent appropriation to benefit national development. The means is by government setting the terms for exploration, production, transportation, and distribution of energy resources.

A resource-nationalist orientation might not cause disputes between governments and foreign investors, but it certainly makes existing tensions more difficult to resolve. Politicians can mobilize the grievances relatively easily through resource-nationalist rhetoric, and this language can be used to attack political opponents or trying to improve the government's bargaining position with foreign investors. Sudden changes in the terms of investment project without consultation will undermine trust between host governments and foreign investors. Populist cries for governments to exercise sovereign power decisively also make it harder for state officials to work toward a mutually acceptable compromise. As such, resource nationalism leads to a greater number of arbitrations and project cancellations.

On the demand side, energy resource nationalism leads to nationalist and mercantilist strategies to seek energy resources and commodity supplies, employing tactics ranging from assertive diplomatic and financial support for acquiring oil and gas reserves to using military posturing and action to secure resource deposits and protect supply lines such as sea lanes.[5] In Asia, over the past decade, state-backed national oil companies (NOCs) have become increasingly competitive with international oil companies, with Chinese NOCs representing the largest winning group in the bidding for Iraqi oil field development. India's Oil and Natural Gas Corporation and others have become significant oil and gas investors in the South China Sea since recent years. Japan and South Korea also have sought to reinvigorate their earlier NOC drive abroad.

In fact, resource security strategies are not without historical precedent in Northeast Asia. During the 1960s and 1970s, the Japanese government offered financial and diplomatic assistance to its industrial corporations to sponsor the development of new mining firms in Latin America and Southeast Asia. These mining firms were connected to the Japanese economy through long-term contracts and investment ties, which guaranteed Japan's resource security by ensuring that their output was preferentially supplied to the Japanese market.[6] Korean heavy industrialization in the 1970s also saw its government launch a similar resource diplomacy programme to strengthen intergovernmental ties with its foreign energy suppliers.[7]

These two countries have recently launched new resource security strategies and sped up their overseas investment in resource sectors. For example, from 2006–10, the average annual growth rate of Korean overseas resource investment was 50 per cent, and the percentage of resource investment in its total outward foreign direct investment (OFDI) increased from 2.5 per cent in 2000 to 31.5 per cent in 2010. By 2010, the state-owned enterprises (SOEs) accounted for 75.1 per cent in Korean total overseas resource investment, 61.1 per cent in its overseas oil and gas investment, and 13.7 per cent of its total investment in other mineral resources.[8] From 2004–10, Korean government had passed and implemented four Overseas Resource Development Master Plans, aiming to enhance self-development rate of overseas mineral resources and increase its energy resource security. Similar to China's programme, Japan and Korean strategies have also called for national firms to make security-oriented investments in overseas resource projects, providing government assistance to national firms to acquire ownership, and ultimately control, of overseas resource projects. But little attention has been paid to the strategies of these two countries.

On the supply side, resource nationalism has also increased in Southeast Asian countries like Myanmar, Indonesia, and Vietnam, due in part to a surge in global energy demand particularly in two Asian economic giants — China and India. Consequently, foreign investors in natural resources face increasingly potential risks in the region. In one 2012 resource nationalism index, Myanmar and Indonesia ranked fifth and sixty-ninth respectively of 197 countries in the "high risk" category.[9] The other seven countries in the "extreme" resource nationalism risks by the index were Somalia (ranked highest at 1), Democratic Republic Congo (2), Sudan (3), Turkmenistan (4), Yemen (6), Iran (7), and Guinea (8).

CHINA'S ENERGY DEVELOPMENT IN MYANMAR

The Importance of Myanmar

For China, Myanmar is geopolitically significant and important given its access to the Indian Ocean and the Andaman Sea, especially at a time when China's long-standing ally, Pakistan, has been struggling to contain Islamic extremism and domestic political unrest, and India, China's potential competitor, is working hard to push forward its "eastward expansion". Myanmar's strategic location in Southeast Asia, connecting India and China, and serving as a crucial link in the Asian/ASEAN Highway Networks, Singapore-Kunming Rail Link, and Trans-Asia Railway Network, highlights the importance of Myanmar as a key node in connecting global trade.

Before the State Law and Order Restoration Council (SLORC) took power in September 1988, all Myanmar governments had prohibited foreign participation in oil and gas exploration and production. In 1988, the SLORC opened up the opportunity for foreign companies to explore for oil and gas, and its gas production increased from 3.4 bcm (billion cubic metres) to 12.7 bcm in 2012.[10] Myanmar's current contribution to the region's gas supply is relatively modest. Its total gas exports of 9.9 bcm in 2007 were less than a third of either Indonesia's (33.1 bcm) or Malaysia's (31.6 bcm). These natural gas exports are currently produced from the offshore Yadana and Yetagun fields in the Gulf of Martaban, and considerable untapped exploration potential is believed to exist on the basis of the large gas volumes being extracted in the Yadana and Yetagun fields.[11] Additional production in the Bay of Bengal, including from the prospective Shwe fields, is set to come on stream soon.

Though Myanmar is not a major energy supplier to China, Chinese NOCs have demonstrated increasing interest and activities in the country's energy resource development. China National Petroleum Corporation (CNPC), China Petroleum Chemical Corporation (Sinopec) and China National Offshore Oil Corporation (CNOOC) have all started oil exploration projects there, while China Power Investment Corporation (CPI) has invested in hydropower. The most significant one is the China–Myanmar pipelines project which has strategic and regional implications.

China–Myanmar Oil and Gas Pipelines Project

The China–Myanmar pipeline project comprises multiple separate projects, each with distinct contract ownership structures. The major components are a deep-water natural gas development project and onshore gas terminal; an onshore natural gas transport pipeline from western Myanmar to China; and an onshore oil transport pipeline from western Myanmar to China.

Onshore Shwe Gas Pipeline

The overland Shwe gas pipeline begins at the offshore pipeline natural gas terminal and runs from Kyaukphu to Muse in Myanmar before entering China at the border city of Ruili in Yunnan Province. The gas pipeline became fully operation in October 2013 with the completion of the final Chinese section. The 793-kilometre-long Myanmar section was completed in the first half of 2013 at a cost of approximately US$1.04 billion and became operational in July, notwithstanding protests from environmentalists and local groups affected by the pipeline. South-East Asia Pipeline Company Limited (SEAP), a Hong Kong-registered entity created by CNPC, and the Shwe Consortium members are in charge of constructing and operating this onshore pipeline.

Onshore Crude Oil Pipeline

For almost the entire distance across Myanmar, a crude oil pipeline runs parallel to the gas pipeline. The crude oil pipeline is 771 km and stretches its way into Yunnan and eventually to Chongqing in China. China's CNPC built the pipeline which will transport oil from the Middle East and Africa to southwestern China. The project also involves construction of a new deep-water crude unloading port and oil storage facilities on Myanmar's Maday Island.[12] CNPC controls a 50.9 per cent of stake in the oil pipeline through its wholly owned subsidiary SEACOP (South-East Asia Crude Oil Pipeline). Myanmar's state-owned MOGE (Myanmar Oil and Gas Enterprise) controls the remaining 49.1 per cent.[13] SEACOP is responsible for the construction and operation of the pipeline, while Myanmar's government provides security for the pipeline.

Broader Objectives

Beyond its clear energy strategic value, this oil and gas pipeline project creates other opportunities for economic cooperation and integration between China and Myanmar, and Southeast Asia. The China–Myanmar oil and gas pipeline project is to link up the logistics system between China and Myanmar. The states and provinces in Myanmar the route will pass include Rakhine, Ayeyarwady, Bago, Magway, Mandalay, and Shan state. Once completed, a prosperous economic corridor is expected to be formed along this route, and this corridor will strengthen China's already substantial economic ties with Myanmar and extend its economic influence to Southeast Asia.

In China's consideration, the pipelines will bring development opportunities to its economically underdeveloped south-western provinces, in particular Yunnan and Sichuan provinces. China's booming economic growth in the past decades has created lopsided economic development on the national level. Much of the interior has been left far behind and is vastly underserved by internal gas and oil distribution networks. Currently, south-east China is the only region in the country that lacks oil refineries. As such, it imports oil products from distant refineries in Guangdong and Gansu Provinces through the Maoming–Kunming and Lanzhou–Chengdu–Chongqing oil product pipelines. It is expected that the pipelines would not only alleviate Yunnan Province's oil shortages but also would diversify the province's economic structures.

While the crude pipeline is to be extended to Kunming, a refinery and ethylene plant with an annual capacity of 20 million tons and 1 million tons respectively will also be built there.[14] Using the crude oil from the Myanmar pipeline, this Yunnan refinery would meet the majority of demand from south-west China for oil products. With more crude oil delivered through the pipeline in the future, the refinery is intended to be sufficient to provide oil products to other provinces and even Southeast Asia. More importantly, the building of the refinery and ethylene plant would help Yunnan to diversity its traditional economic structures and turn the province into the country's petroleum and petrochemical base in the south-west. This can help Yunnan Province receive favourable policy support from the central government and attract more investments from Chinese SOEs and big international multilateral companies to build its petroleum and related sectors, thus earning more opportunities to further open up its economy. Hence, the Province would eventually assume the strategic importance as China's south-west "bridge-head" to connect Southeast Asia and South Asia for movement of capital, goods, and labour services.[15]

Myitsone Hydropower Dam Project and Letpadaung Copper Mine

Myitsone Hydropower Dam

The Myitsone hydropower dam is a large dam, suspended due to nationwide protests since 30 September 2011, on the Irrawaddy River in Myanmar. If completed, it will be the fifteenth largest dam in the world. It is planned to be 1,310 m long and 136.9 m wide. The dam project is a joint venture between China Power Investment Corporation (CPC) and Asia World (Myanmar), a military-linked company. It is estimated that the cost of the project will be US$3.6 billion and will provide between 3,600 to 6,000 megawatts of electricity (90 per cent to Yunnan, 10 per cent to Myanmar).

The project has been controversial because of the flooding area, environmental impact, livelihood costs to local people, and the historical significance of its position. More importantly, the main developer of the project, CPI, held negotiations with the Myanmar military government rather than with the local Kachin Independence Organization. Consequently, as government troops entered the territory to secure the project, tensions were increased and conflicts were renewed between Kachin and the government troops, and this became one of the several factors that led to the ceasefire breakdown.

Letpadaung Copper Mine

The Letpadaung copper mine is located near Monywa in Sagaing Division. It is open-pit copper mine 110 km west of Mandalay. It has been mined since 1998, first by Ivanhoe, a Canadian company, and since 2011 by Wanbao Mining (a subsidiary of state-owned Chinese arms manufacturer — China North Industries Corp.) in a joint venture with Union of Myanmar Economic Holdings Ltd., a Myanmar military-linked company. The Letpadaung deposit has 1.478 billion tons of copper ore with a copper content of 0.38 per cent, as well as about 577 million tons with 0.44pc copper, according to 2007 estimates, making it the largest deposit in Southeast Asia.[16]

But this China-backed copper mine became another controversial project when it was planned to be expanded in 2011. Local villagers said that more than 7,800 acres (3,156 hectares) of land from twenty-six surrounding villages have been confiscated to allow for the expansion of the mine, and the confiscations have given rise to a series of local protests.[17] While environmentalists say that the Letpadaung copper mine is in a watershed area near the confluence of the two most important rivers in Myanmar, the Irrawaddy and the Chindwin, around 15 miles west of Monywa,[18] local people have voiced concern over mine pollution contaminating water supply.

As a result of the growing tension over copper mine expansion and the demonstration on 29 November 2012 in which more than a hundred people were injured after a police crackdown, Myanmar President Thein Sein initiated a Commission of Inquiry, approved by parliament, and headed by Aung San Suu Kyi. On 12 March 2013, the Commission of Inquiry released its long report.

In the seventy-four-page report, the Commission recognizes that the project failed to meet international environmental and social impact standards or adequately compensate villagers for their land. It goes on to recommend greater transparency in compensation negotiations and for reclaiming other land for compensation.[19] The report also acknowledges that the police used smoke bombs containing phosphorous against protesters, burning a hundred people, conflicting with earlier official reports. However, the report does not recommend the project end or be suspended. Instead, it gives suggestions for meeting international standards. The local community activists did not accept the report as it does not adequately address their concerns. The Letpadaung copper mine incident reflects a fact that there is a balancing process between agreements signed under the previous regime and the concerns of the local communities.

CAN ENERGY COOPERATION STRENGTHEN CHINA–MYANMAR RELATIONS?

China's involvement and recent flaws and risks in Myanmar have generated a lot of debate and different speculations that Myanmar is moving from China's "backyard" to "competition frontier", or Myanmar is "defecting" from China. However, from both the perspectives of China and Myanmar, the bilateral relationship is based on pragmatism.

From China's Perspective

Myanmar's good relations with China can be traced back to the late 1940s. After Myanmar achieved its independence in 1948, it recognized the establishment of the People's Republic of China in 1949. In Myanmar's view, it seems that China had always regarded Myanmar as essential to its security as Myanmar "stands high in the degree of importance China attaches to its peripheral areas",[20] and the two countries established diplomatic relations in 1950. For the past six decades, China–Myanmar relations have been premised upon the five principles of peaceful coexistence, agreed upon by Myanmar, China, and India in 1954.

However, bilateral relations had been undermined by the incursion of Chinese troops into the northern Shan State and Chinese support of the Burma Communist Party (BCP) rebels in the 1950s,[21] and reached the lowest point of all time when "Mao's 'Culture Revolution' was exported to Myanmar's overseas Chinese community and led to violent riots in the capital city in June 1967."[22] Myanmar recalled its ambassador and students from China while Chinese technicians working on technical assistance projects were expelled from Myanmar.

When the United States and some Western countries imposed political and economic sanctions on Myanmar in the late 1980s, pressing the military regime on human rights issues, democratization, and the release of political dissidents, China took the advantage of this situation to recover and cultivate closer relations with the military government through trade, investment, and large-scale infrastructure projects undertaken by major Chinese to build dams and transport natural resources. China's economic and strategic interests, as well as political clout, have steadily risen in Myanmar since Beijing reversed previous policies and withdrew support from the insurgent BCP in the late 1980s. This contributed to the BCP's later collapse through a mutiny in 1989, and in its splintering to the formation of several ethnic-based insurgent organizations now active along the China–Myanmar border.

In recent years, with its financial power increasing, China's advantage is that it has become one of the major investors in Southeast Asia. As analysed in Chapter 3, China's FDI in ASEAN had increased from US$195.7 million in 2004 to US$7.3 billion in 2013, and China's FDI in Myanmar increased from US$4.1 million to US$475 million. However, although China has planned to stake a long-term strategic energy investment in the country and its capital has largely poured into resource and infrastructural sectors, it does not necessarily mean that China understands Myanmar's local societies and the bilateral relations can be strengthened. China's experiences show that although many Myanmar watchers believe that Myanmar has long been one of "China's few loyal friends", this is no longer true. Just a few months after Myanmar's new president Thein Sein took power, Naypyidaw successfully challenged its long-term friend by suspending construction of the Myitsone dam in September 2011 on the reasons of environmental protection. When reviewing the suspension of the project in 2012, CPI President Lu Qizhou concluded that "as a central state-owned enterprise, the CPI has not been used to dealing with non-governmental organizations and the local community. This is a lesson that central SOEs need to learn when 'going out'".[23]

Indeed, in the current context, where ASEAN member nations are looking toward economic integration, the role of the environment within the economy has received increased global attention. In view that ethnic groups and NGOs (non-government organizations) have potentially a great impact on government policies, China needs to make its approach more transparent with respect to its intentions and stance toward domestic issues in Myanmar.[24]

From Myanmar's Perspective

Myanmar is the second largest country in ASEAN and is situated in the interface of South and Southeast Asia. Sandwiched between the two most populous countries in the world — China and India — Myanmar has always been conscious of the geographic and demographic realties in formulating its foreign policy. The fact that this country is inhabited by over one hundred indigenous nationalities with many of them living along the borders also complicates Myanmar's policy that has to take into consideration the dynamics of the international actors and domestic economic, political, and security concerns. Over the past years, China has expended considerable effort to improve its neighbourhood relations, with the goal of maintaining stable borders and a viable trading environment. China has, in the past, described its interests in Myanmar as stability, border security, energy security, and connecting landlocked Yunnan Province to neighbouring markets. In pursuing these interests, it has undertaken significant diplomatic efforts in maintaining dynamic networks among governments and governing elites.

However, as time passes, one of the main shortcomings of this policy approach is becoming evident — China is increasingly finding itself at odds with non-governmental actors and local people, and it has become difficult to deal with increasingly negative perceptions across Myanmar society. In Myanmar, most of the dams are as Chinese as the dams in China. They were built either through exclusive concessions to Chinese firms or by China–Myanmar joint ventures where China provides the engineering technology and the loans.[25] The Myitsone Dam in Kachin State had commissioned both Chinese and Myanmar biodiversity experts, but there were no consultations with local stakeholders, who are more concerned about how the project will affect their interest. As they know that large-scale water and mining projects usually require that people move away, often without either national or human security.

More importantly, China's investment projects in ethnic areas of Myanmar, and Kachin State in particular, are increasingly touching on fundamental questions of political self-determination and national conciliation. With the country in the midst of a tentative national ceasefire process, land acquisition has become a key driver for protests, and dam sites have the potential to stir latent or resurrect suspended conflicts, as some of the construction sites are often within or nearby recent conflict zones.[26] The Burma River Network, and alliance of several local environmental activist groups, believe that up to fifty clashes between military forces and armed ethnic groups have occurred in connection with ongoing hydropower projects in the last three years.[27] While this cannot be independently verified, it is nevertheless apparent that the ongoing peace process needs to address land rights and other issues that are directly associated with dam construction. As such, dam construction remains a bargaining chip for ethnic groups to use at the political process and negotiating table.

In Myanmar's view, China's half-hearted engagements at the local level and poor crisis management have added to the widespread perception that China is solely concerned about the security of its own business operations and energy security, while ignoring the needs and interests of ethnic nationalities.[28] Indeed, Chinese investment projects, primarily in energy and natural resources, have failed to create sizeable employment opportunities for the locals. At the same time, the military government of Myanmar did not spend the earning from deals with China on improving the people's livelihood.[29] As such, China's presence in Myanmar has not brought substantial benefits and has instead strengthened the locals' perception that the economic relations between the two countries were unequal. In particular, locals believe that China is grabbing their natural resources and disregarding their interests.[30]

JAPAN AND INDIA AS IMPORTANT PLAYERS

As Myanmar opens up and its domestic reforms are going on, Myanmar is likely to see its old friend stream in. Japan and India can be expected to step up their engagement. The entry or resurgence of more global players like Japan and India will doubtless provide Myanmar with more options, facilitating its attempts at loosing itself from the Chinese grip. But on the other hand, this has also prompted China to realize that it needs to adjust its investment schemes and redefine its resource diplomacy in Myanmar and other Southeast Asian countries so as to better embrace the new environment and dynamic changes in the region.

Japan's Return to Myanmar

In the 1950s, Japan had emerged as a key player in Myanmar, leveraging its aid programme to secure political influence and economic access. Japan's economic assistance began in the form of war reparations. In November 1954, Japan and Myanmar reached an agreement on reparations of US$200 million to be paid out over ten years, as well as US$5 million annually in Japanese goods and services for joint projects. During the Ne Win era, almost 40 per cent of Myanmar's imports came from Japan, and from 1976 to 1990, more than 66 per cent of the bilateral official development assistance (ODA) Myanmar received came from Japan. After the military coup in 1988, however, Japan–Myanmar relations changed markedly. This was particularly reflected in the sharp decline of Japan's aid to Myanmar. As domestic and international pressure mounted for Japan to support international sanctions on Myanmar, Japan had to reduce its aid to Myanmar. Annual average aid from Japan to Myanmar was US$155 million per year between 1978 and 1988, but during the period from 1996 to 2005, that figure was down to an annual average of just over US$36 million.[31] As Japan gradually pulled back from Myanmar, China's influence deepened and took over from Japan as the key supporter and backer of Myanmar's military rulers in international affairs, and gradually became the biggest trade partner and investor of Myanmar.

With the return of Prime Minister Shinzo Abe to power in December 2012, Tokyo has adjusted its Southeast Asia policy, and Japan has begun a new relationship with Myanmar. In May 2013 Prime Minister Abe paid an official visit to Myanmar, which was the first visit by a Japanese prime minister since 1977, symbolizing Japan's return to Myanmar after a long hiatus.[32] The two countries issued a joint statement in a "cordial and friendly atmosphere". PM Abe not only wrote off billions of dollars-worth of Myanmar's debts, but also pledged to implement assistance totalling 91 billion yen (US$1.24 billion) to Myanmar by the end of the current fiscal year.[33] Some of it will go to poverty reduction, power rehabilitation and infrastructure development. Once again, Tokyo used its aid programme to signal its enthusiasm in broader economic and diplomatic engagement with Myanmar. PM Abe also met Aung San Suu Kyi and promised "all possible assistance" to support the country's commitment to democratic reform.

Economically, in December 2012 Japan signed a contract with Myanmar that allowed Japanese banks and companies to enter Myanmar. At least thirty-five Japanese investment projects are underway in Myanmar, the biggest being plans to develop the 2,400-hectare Thilawa Special Economic Zone near Yangon, led by trading companies Mitsubishi, Marubeni Corp, and Sumitomo Corp.[34] Thilawa is the flagship of Tokyo's commitment to promote industrialization, employment, and economic development in Myanmar, as Tokyo anticipates that the success of the Thilawa industrial park will boost local economic modernization and political stability. Another major development is the Dawi (Tavoy) port and industrial area near Thailand. Japanese aid is providing a road from Bangkok, and the private sector is supposed to develop this massive industrial area with a total of some US$13 billion over the coming years. This port, according to Japanese authorities, will enable Japan to have direct access across the Bay of Bengal to Chennai (Madras) in India, and stimulate an increase in Japanese investment there and beyond.

As China's diplomatic frustration mounted, Japan seized upon domestic shifts within Myanmar to restore its economic engagement with the country by increasing its ODA and investments. Enhancing engagement with Myanmar is part of Tokyo's Southeast Asian strategy. In PM Abe's consideration, by turning to Southeast Asia and Myanmar in particular,

Japan's security strategy toward China would be moving away from a narrow focus on a potential military conflict in the East China Sea to a broader regional and international context.[35]

Competing for Good Relations with Myanmar

Trade and investment

Before 1988, Japan was Myanmar's major trade partners due to aid-driven trade. As Myanmar's trade volume grew, its geographical trade patterns changed. Myanmar has strengthened its trade relations with its neighbouring countries, especially China. After 1990, China became the most important trade partner to Myanmar and replaced Japan as a supply source of imports for Myanmar. While Japan–Myanmar bilateral trade increased just 3.4 times, from US$324 million in 2000 to US$1.1 billion in 2011, China–Myanmar bilateral trade increased 10.4 times, from US$659.5 million to US$6.8 billion.[36] This was also because of Japan's suspension of its ODA and the supplies related to its economic cooperation programmes decreased sharply. Figures show that in 1987, Japan constituted nearly 40 per cent of Myanmar's total imports, while China accounted for just 3 per cent.[37] After 1990, China's share of Myanmar's imports continued to rise, reaching nearly 40 per cent in 2011; while that of Japan dramatically declined to less than 7 per cent in 2011.

In terms of investment, according to ASEAN statistics, from 2002–9, China's accumulated FDI in Myanmar amounted to US$953 million, accounting for 26 per cent of the total FDI inflows into Myanmar, while Japan's accumulated FDI in Myanmar during the same period was only US$8.5 million, accounting for just 0.23 per cent of the total FDI inflows into Myanmar.

China's most striking adjustment toward Myanmar is the drastic shrinking of Chinese FDI since in 2010. According to China's statistics, between 2007 and 2010, Chinese FDI in Myanmar jumped from US$92.3 million to US$875.6 million, and decreased to US$217.8 million in 2011.[38] The boost came primarily in 2010 — the Myitsone dam project, the CNPC pipelines, and Norinco's Mongywa copper mine project valued collectively more than US$8 billion. However, since the political reforms in Myanmar picked up speed, this rapid growth was abruptly interrupted due to various reasons. But in any case, considering China's overwhelming economic importance to the country, it will be hard for Myanmar to find a realistic alternative to China to meet its economic needs.

ODA

Japan's ODA programme originated out of the provision of reparations to Asian countries following the San Francisco Treaty of 1951, and subsequent bilateral agreement reached with Myanmar (1955), the Philippines (1956), and Indonesia (1958). The fact that the foreign aid was linked closely to reparation had some important long-term effects on Japan's ODA programme. First, Japan was reluctant to impose political conditions upon its aid recipients. This was evident in Japanese aid to China, but was also characterized in Japanese aid to Myanmar and Indonesia. As late as 1990, Japanese Ministry of Foreign Affairs explicitly equated such conditions with engaging in undue interference in the internal affairs of recipient governments.[39] Second, Japanese aid has traditionally been oriented toward its Asian neighbours. Up until 1972, some 98 per cent of all Japanese aid was directed to the Asia Pacific. By 1990, Asia still received 59 per cent of all Japanese aid. Japanese ODA was explicitly regarded as a legitimate arm of national policy.[40] Third, the strategic use of aid was particularly evident in Japan's resource diplomacy of the 1970s, which provided aid and investment to countries possessing resources which were essential to Japanese domestic industrialization process.

Myanmar was the first recipient of Japanese post-war reparations. Throughout Ne Win's sixteen years in power (1962–88), Myanmar remained one of Japan's largest aid recipients, receiving a total of US$2.2 billion. By 1987, Japan's ODA constituted 20 per cent of Myanmar's national budget, making up 71.5 per cent of the total foreign aid that Myanmar received.[41] Japan adjusted its ODA policy to Myanmar following the military coup in March 1988, and the amount of its aid declined significantly. For example, Japanese aid was provided to Myanmar at the average annual amount of US$154.8 million for the period 1978–88. This declined to US$86.6 million for the period 1989–95, and then to US$36.7 million for the period 1996–2005.[42]

From the early 1990s, Japan started to change its ODA policy. In 1992, Japan adopted its first ODA Charter which placed greater emphasis on human rights and democracy, shifting from pure economic interests to the development of human capital, values, and political systems in host countries. Japan, especially in its ODA motivations, emphasizes that "it was a recipient of economic aid before it attained the status of an economic superpower" so that it knows the importance of aid-giving from the perspective of recipient countries.[43] Thus, "it may be able to serve as a role model for other developing countries and to be less driven by self-interest."[44]

Japan wants to support Myanmar in human resource development, and provide technical assistance and financial support for the development of the small- and medium-sized enterprise sector. In addition, it has provided humanitarian assistance through Japan's Nippon Foundation, which was allowed to deliver US$64,000 of aid directly to internally displaced persons, including rice and medicines.[45] The Nippon Foundation has adopted the framework of "human security" to provide humanitarian assistance and is working with the government-affiliated Myanmar Peace Center to provide aid to various ethnic minority groups as it has invited key members of the United Nationalities Federal Council (a loose coalition of Myanmar's armed ethnic minorities) to meet in Tokyo to discuss humanitarian assistance to their regions. Moreover, Japan has sought to embed ODA to Myanmar within a regional framework oriented around Mekong River countries. In 2011, Japan's grant aid and new loans to Cambodia, Lao PDR, Myanmar and Vietnam (CLMV) accounted for 44.5 per cent of its total grant aid and new loans to ASEAN countries, with technical cooperation accounting for a large part of Japan's ODA.[46] In particular, providing support for the Mekong countries is important from the standpoint of reducing intra-regional disparities and poverty.

China is one of the world's most experienced providers of ODA, with an aid programme dating back to 1950. But its aid programme was mainly expanded in support of new nations "fighting against colonialism and hegemony" as it found itself at loggerheads with both the Soviet Union and the United States by the 1960s. In the mid-1990s, China instituted a series of reforms in its aid programme, explicitly linking its aid to benefits for China as much as it is to benefits for recipients.

China began its ODA to Myanmar in the late 1980s when Japan suspended its ODA to Myanmar in 1988. In 1991, China pledged its first major aid grant of US$8.6 million to Myanmar and expand its aid programme in concert with its strategic investments. In July 1993, China promised an interest-free loan of US$8.6 million, and the two sides celebrating the opening of the Yangon-Thanlyin Bridge, supported by loans of US$29.1 million. Between 1997 and 2006, China provided some US$24.2 million in grants, US$482.7 million in subsidized loans, and $1.2 million in debt relief.[47] China also supported a number of build, operate, and transfer (BOT) agreements, including the Hmawbi Rubber Ball Factory, Tyre Factory, Belin Sugar Mill, Shwedaung Textile Mill, and Meikhtila Textile Mile. China has also expanded its support for infrastructure projects to Myanmar, strengthening boat, rail, and road links that would enable goods to be shipped in by sea to container ports in Myanmar, offloaded, and then taken by rail and river barge over to the Chinese border. China's ODA in Myanmar is closely linked to China's own economic and strategic interests, and shows no signs of even nascent socialization.[48]

Like Japan in the mid-1970s, China was facing an energy shortage by the early 1990s as it became a net importer of petroleum in 1993. China began to leverage burgeoning capital resources to secure access to strategic natural resources, with its first subsidized loan to Sudan in 1996 to finance oil exploration through a joint venture with China National Offshore Oil Companies. Most of its resource-related loans are given on a commercial basis, and would likely not qualify as ODA under OECD (Organisation for Economic Co-operation and Development) standards.[49] According to China's 2011 white paper on foreign aid, China had provided US$39 billion in aid to foreign countries by the end of 2009, including US$16.6 billion in grants, US$11.6 billion in interest-free loans, and US$11.19 billion in concessional loans, with Africa and Asia accounting for 46 per cent and 33 per cent of China's foreign aid respectively.[50] Although the white paper does not break the data down by country, it is obvious that the amount of trade, investment and aid directed to Southeast Asia — in particular via infrastructure financing — has grown substantially in recent years, and that China is one of the largest sources of economic assistance, defined broadly, in Southeast Asia,[51] with Myanmar being one of the major recipients.

Like Japan's earlier aid programmes, Chinese ODA and subsidized loans are often part of a larger investment package aimed at securing access to key strategic energy resources. China's investment and economic cooperation in Myanmar and other Southeast Asian countries, is mainly motivated by two objectives — to maintain a favourable neighbouring environment, and to secure natural resources, energy resources in particular. Both objectives are critically important to China's economy and to its aim of becoming a global economic power. China's Export-Import Bank has described ODA as a "vanguard" supporting Chinese exports and investments while contributing to sectors such as transportation, telecommunications, and energy, "thus improving the investment environment in developing countries".[52]

China also claims that it is in a position to understand the concerns of developing nations. China has historically maintained the line that it is, and will remain for a long time yet, a developing nation and has thus implied that it shares the issues and problems of many less developed countries.[53] While acknowledging that China is engaging in "resource diplomacy", Beijing maintains that "in growing relations with our neighbors and other developing countries that have long been friendly toward China yet face daunting challenges in development, we will accommodate their interests rather than [be] seeking benefits at their expense or shifting troubles unto them".[54] However, the prevailing view in Southeast Asia is that, "in the eagerness to deploy Chinese capital and expertise for rapid completion of resource extraction, transportation and power-generation projects, Chinese SOEs (state owned enterprises) have been given wide leeway in disregarding environmental standards and the interests of local people affected by these projects".[55] Many outside observers are concerned that China will also replicate the sort of "neo-mercantilist" strategies that Japan adopted during its high-growth phase in the 1970s, a possibility that is reinforced by the prominence of "state capitalism" in a number of rising powers.[56]

Win-win Competition in Myanmar?

Both China and Japan view Myanmar as a country of strategic importance and as a place where huge economic profits can be made. Since China has a twenty-year head start, it is unlikely that Japan can match the Chinese position in Myanmar. More importantly, China has a geographical advantage over Japan as its Yunnan province is adjacent to Myanmar and it is building a deep water port at Kyaukpyu in the Bay of Bengal and an extensive network of roads to Yunnan. However, Japan still remains the world's third largest economic power, with a growing soft power base, especially in Southeast Asia where its pop culture is widely enjoyed. "Millions of Burmese consumers still prefer high-quality Japanese cars and electronics over cheaper Chinese products".[57] The Japanese language remains one of the most popular languages for Myanmar people to learn. Moreover, Japan attempts to internalize and apply "Western standards" (such as respect for human rights, liberal democracy, and free-market based economy) to its relations with Myanmar. The resurgence of Japanese interest and influence in Myanmar, together with the positive roles being adopted by the United States, India, and the European Union, may mitigate China's prominence in the country and help Myanmar resume the neutralist stance it held in the Cold War.[58]

However, Myanmar's foreign relations need not constitute a zero-sum competition between China and Japan. The two countries offer Myanmar different types of engagement. Japan's investment is more focused on manufacturing (Southeast Asia accounted for less than 2 per cent of Japanese mining-related FDI in 2010), and Tokyo is more willing than China to work with civil society organizations in areas such as human rights and development, while China is more interested in energy and natural resources. A positive sign is that China, in its activities in Myanmar, has lately been applying what it had learnt from Japan's practices over the years in Southeast Asia in humanitarian assistance. Beijing seems to be realizing that it needs, like Japan, to adjust its ODA and investment schemes and redefine its political, economic and humanitarian role in Myanmar to function well in the new environment. More concretely, China needs to take positive steps such as working with non-profit or non-government organizations, just like Japan and other Western countries do, to effectively help the poor in specific areas.

Indeed, after the suspension of the Myitsone dam project in late 2011, Beijing has made corporate social responsibility programmes an integral component of Chinese SOEs' operations in Myanmar. China has recently initiated what it calls a "people to people" relationship with Myanmar. It has arranged several friendship tours targeting various political parties in Myanmar, including the National League for Democracy (NLD), civil society organizations such as the 88 Generation Peace and Open Society, as well as local media groups to help build a better understanding between the societies of the two countries. All these steps are quite unprecedented where Chinese policy towards Myanmar is concerned. If Japan's "return" to Southeast Asia can indeed prompt China's economic and resource diplomacy in Myanmar and other Southeast Asian countries in this positive direction, the implications may be significant. For one thing, it would mean that China's economic approach would become less crude and more charm-oriented.

India's Engagement with Myanmar

As a neighbour to India, Myanmar's location is central to strengthening India's "look-east policy", energy security, and counterbalancing China's growing influence in Southeast Asia. In the face of China's rise and its increasing influence in Southeast Asia, the major concern of India is that the close relations between Myanmar and China might change Myanmar's traditional policy of neutrality.

Strategically, India is afraid that China will attempt to form a "strategic encirclement" against India, using Myanmar as a point to contain India.[59] The Indian government was particularly worried about the

China–Myanmar strategic links and the prospects of the Chinese Navy gaining a foothold in the Bay of Bengal. India sees China, which has close relations with Pakistan and Myanmar, as a potential threat. China's involvement in Myanmar could mean that India would be surrounded on three sides by Beijing and its area of influence, leaving no buffer states. India feels that it badly needs to improve its strategic relations with Myanmar so as to break down China's encircling strategy.[60]

Another concern is the energy security. Energy has become a bottleneck restricting India's economic development. India has 0.5 per cent of world's proven oil reserves, its oil production accounted for only 0.9 per cent of world oil production, but it consumed 3 per cent of the world's total oil consumption. India also has limited gas resources, it has 0.6 per cent of world's gas reserve, and its gas production accounted for 1.1 per cent of world's total production, while its consumption accounted for 1.3 per cent.[61] As a net gas and oil importer with an ever-increasing domestic demand, India looked to gain direct access to Myanmar's abundant natural gas reserves. These could potentially be accessed more cheaply and securely than alternative pipeline routes from Central Asia that would need to travel through Iran, or Afghanistan and Pakistan. Therefore, India has been very concerned over China–Myanmar energy cooperation and the building of China–Myanmar pipelines, especially when Myanmar turned to cooperation with China after its negotiations on several projects with India failed.

India's Interest in Myanmar

Historically, India and Myanmar were part of the British Empire, Burma being the largest province in British India. In 1937, Burma became an independent unit within the empire, and there were 300,000 to 400,000 Indians who had migrated there under the British Empire running the public services, police, and the military. Many of the tribes in India's north-eastern region are ethnically linked to tribes on the Myanmar side of the border. Currently, at least five major militant groups from India's north-east, where numerous tribal and ethnic groups are fighting for greater autonomy and fuelling violent insurgencies, have training camps in the dense jungles of Sagaing in northern Myanmar.

India–Myanmar relations were close during the early years after independence, with both Jawahahal Nehru and U Nu originating and leading the Non-Aligned Movement. Both countries signed a Treaty of Friendship for duration of five years, which was to remain in force "forever thereafter" if neither side gave notice of its desire to terminate it six months prior to its expiration. However, after the 1962 coup, relations came to a standstill. In 1989, when the military junta took over in Myanmar, placing Aung San Suu Kyi under house arrest, India imposed financial and trade restrictions on Myanmar. The Indian foreign ministry "urged the military authorities to release opposition leaders and create conditions for holding free and fair elections as soon as possible".[62]

India's interest in Myanmar began to recover from the early 1990s when Narasimha Rao assumed power, concurrent with a shift from a low-key policy which emphasized human rights and democracy to one which reflected a pragmatic strategic policy towards Myanmar. In 1993, India shifted its attitude and moved from voicing its opposition to the military junta's crackdown on pro-democracy activists and the arrest of Aung San Suu Kyi to a more pragmatic, non-interventionist policy. In January 2007, Indian External Affairs Minister Pranab Mukherjee said that New Delhi had to deal with other governments "as they exist... We are not interested in exporting our own ideology" and called "democracy and human rights an internal Myanmar issue."[63]

This change in policy by India has been largely prompted by its desire to counter Indian insurgent groups operating from Myanmar. Over the past three years, New Delhi-Yangon counterterrorism initiatives have gained in momentum, and transfers of military equipment have increased significantly.[64] It is likely that joint military initiatives in the border region will be initiated and more direct military aid like the proposed light attack helicopter sales from India to Myanmar will continue. It is predicted that India will continue these operations as part of its efforts to deepen bilateral ties with the military junta in Myanmar.

India has increased interest in investing in energy resources. Indian companies have bid on a number of gas blocks off Myanmar's Rakhine coast. In early 2005, Myanmar, India, and Bangladesh signed a memorandum of intent for the construction of a gas pipeline connecting gas from these blocks to Kolkata via Bangladesh. However, negotiations between India and Bangladesh broke down and in December 2005, Myanmar instead committed gas from these fields to PetroChina under a thirty-year agreement in which gas would be transported to Kunming via a new pipeline. In June 2013, Myanmar resuscitated India's hopes with a proposal for a pipeline using the same route as the to-be completed Kaladan Multi-modal Transport project, from Sittwe port to India's north-east.[65] Indian companies have had more recent success in obtaining oil and gas exploration rights in the 2013 government auction of thirty offshore and eighteen onshore blocks.[66]

The developments in the gas field projects of Myanmar have served to highlight the intense energy diplomacy and competition in this region. The government of Myanmar withdrew India's status as preferential buyer on the related blocks of its offshore natural gas fields and declared its intent to sell the gas to PetroChina, suggesting India's failure in its competition with China for the pipelines. By resuscitating India's interests in a pipeline and exploration rights, Myanmar is getting closer to India, illustrating Myanmar's attempts to hedge against China's rising influence and reduce its predominant position in Myanmar's energy projects.

India to Strengthen Its Relations with Myanmar

China and India are emerging economies. To a large extent, their rise depends on economic growth. Both China and India are making global searches for energy as part of their energy strategy. Common needs for energy resources make China–India relations more competitive. Therefore, India has been very concerned over China–Myanmar energy cooperation and the building of China–Myanmar economic corridor, especially when Myanmar turned to cooperation with China after its negotiations on several projects with India failed. India is very clear that it needs to strengthen cooperation with Myanmar for energy resources and market access to Southeast Asia. "India's current setback in the field of energy is unlikely to lead to a decrease in its attempts to compete with China in other fields."[67] India will undoubtedly make more efforts to establish a stronger presence in Myanmar.

India chose to engage Myanmar through subregional organizations such as the Bangladesh–India–Myanmar–Sri Lanka–Thailand economic cooperation and the Mekong–Ganga Cooperation (MGC) which was founded end 2000. These projects underline cooperation in tourism, culture, education, and language training, emphasizing the "natural connectivity" between India and mainland Southeast Asia based on cultural and civilizational similarities. The MGC has not only boosted regional trade, especially between Vietnam and India, but also another forum for Indo-Myanmar interaction.

India has taken some other concrete approaches to strengthen its economic and military relations with Myanmar. From 2003–11, India's exports to Myanmar increased from US$86 million to US$513 million;

imports increased from US$390 million to US$1.2 billion.[68] In terms of investment, as of March 2010, India's cumulative direct investment in Myanmar reached US$20 billion. On energy cooperation with Myanmar, although India failed in its competition with China for the pipelines, its energy cooperation with Myanmar is in continuation. India is the main shareholder of several oil and gas projects which are under construction in Myanmar. India also holds shares in China–Myanmar pipelines project, and is negotiating with Myanmar on building oil and gas pipelines to western and eastern India. In February 2010, the Indian Cabinet Committee on Economic Affairs approved requests by Oil and Natural Gas Corporation (ONGC) and GAIL (India) Limited to invest US$1.35 billion in Myanmar, with the two companies increasing their stakes in Blocks A-1 and A-3 to 20 per cent and 10 per cent respectively, and acquiring 8.4 per cent and 4.1 per cent equity respectively in the Myanmar–China gas pipeline.[69]

In terms of military cooperation, Myanmar allowed the Indian Navy flotilla to berth in Thilawa in 2002 and the following years. Moreover, whereas China has not yet achieved the goal of conducting joint operations with Myanmar, India has successfully conducted India–Myanmar joint naval exercises, held in 2003, 2005, and 2006.[70] The other important fact regarding India–Myanmar military cooperation is related to arms supplies. It is said that for balancing dependency on China, Myanmar has renewed its sources of arms suppliers to include India, in addition to its original suppliers — Russia, Pakistan, Singapore, and Ukraine.[71]

While Chinese projects in Myanmar are consistently falling prey to public disapproval, several infrastructure projects that will connect Myanmar to India's north-eastern states appear to be making progress, even in the face of similar local opposition. For example, the Kaladan Multi-Modal Transit Transport Project, which will connect India's Mizoram State to deep-sea port in Sittwe, is projected to be completed by 2015. The project will expand the capacity of the Sittwe port facility, giving India's north-eastern states access to the harbours in the Bay of Bengal and connecting Myanmar to Kolkata port. Another project, the India–Myanmar–Thailand trilateral highway corridor, is projected to be completed by 2016. In July 2014, the governments of India and Myanmar also pledged to proceed with the creation of a highway bus route that will connect Moreh in India's Manipur state to Mandalay.[72] The Moreh–Mandalay highway will fill a crucial gap in the Asian Highway network as well as connect India's north-eastern states to the East-West Economic Corridor, which connects Mawlamyine in Myanmar to Da Nang port in Vietnam. Thus India will achieve a level of connectivity throughout Southeast Asia similar to that which China has enjoyed for decades.

In a recently released report titled "Transforming Connectivity Corridors between India and Myanmar into Development Corridors", former Indian ambassador to Myanmar V.S. Seshadri announced that a broad rail line that is supposed to be built in Imphal by 2018 could be extended to Moreh and on to Kalay in Myanmar's Sagaing division with international funding, which would be another crucial link in regional connectivity.[73] The report listed several other India-assisted developmental projects in Myanmar, which include industrial training centres in Pakokku and Myingyan and an India–Myanmar Center for Enhancement of information Technology Skills.

It is believed that Myanmar's new preference for India-sponsored infrastructure projects reflected Naypyidaw's interest to play an active role in the U.S. policy of rebalancing strategy in Southeast Asia. In his visit to India in July 2014, U.S. Defence Minister Chuck Hagel said, "The U.S. strongly supports India's growing global influence and military capabilities including its potential as a security provider from the Indian Ocean to the greater Pacific."[74] In Washington's consideration, by forging a stronger alliance with India, it may better influence New Delhi over its Myanmar policy. In Naypyidaw's consideration, as it falls further under India's influence, it also embeds itself in American foreign policy, which is seen as a bulwark against China's strategic ambitions in the region. By challenging China's dominant position, the Myanmar government is opening strategic space to create further competition between India and the United States on the one hand and China on the other, thus affording the Myanmar government a more comfortable degree of leverage and autonomy in the international and regional arena.

CONCLUSION

Given the importance of China's capital, technology, and big market, China will remain as Myanmar's major energy cooperation partner. On the Myanmar side, the democratic position and accompanied economic liberalization is viewed as a huge greenfield opportunity for Chinese companies. McKinsey Global Institute, for example, estimates that Myanmar would need to invest US$320 billion in infrastructure by 2030 if the economy is to achieve a growth potential of 8 per cent per year.[75] More importantly, China has maintained close relations with ASEAN. As ASEAN's importance for international markets increases with various energy challenges emerging in the region (such as increasing reliance on the Middle East for oil supply and climate change), China's deepening relations with ASEAN will be of growing strategic importance. The potential for energy cooperation between China and ASEAN countries and Myanmar in particular is high and important.

However, although the relationship between China and Myanmar has generally proved to be mutually beneficial, concerns and resource nationalism in the local society that may impact the bilateral relationship are prevailing. The extent to which China's energy resource cooperation with Myanmar will develop depends on the efforts of both sides. On the one hand, it depends on whether civil societies and local people in Myanmar, primarily labour unions, organized businesses, and civil society formations, feel that their concerns about China's influence on their political economy are being addressed.

> As politics in Myanmar becomes more open and democratic, foreign investors in the resource sector will need to be more cognizant — because of new legislation or by choice — of the social and environmental impacts of these projects, and to make provision for compensation for affected land, ensure opportunities for local employment, and provide transparency about the royalties paid or revenues shared by the government.[76]

On the other hand, it largely depends on whether the Myanmar government can successfully conclude lasting peace settlements with armed ethnic minority groups. Thus far the signs are not promising, as it is unclear precisely how the central government will deal with demands for greater autonomy within the federal union by many of these groups in a way that will not encourage secessionist trajectories.[77] Moreover, it also depends on whether Myanmar can depoliticize some high-profile Chinese deals and do not exaggerate their public debates about the risks of Chinese investment.

Nevertheless, on the China side, Beijing's promise to make corporate social responsibility programmes an integral component of Chinese SOEs' operation in Myanmar suggest that Beijing is trying to understand the sensitivities and complexities of Southeast Asian countries' internal politics. A positive sign is that China, in its activities in Myanmar, has lately been applying what it had learned from Japan's practices over the years in

Southeast Asia in humanitarian assistance. Beijing seems to be realizing that it needs, like Japan, to adjust its ODA and investment schemes and redefine its political, economic, and humanitarian role in Myanmar to function well in the new environment. More concretely, China is taking positive steps such as working with non-profit or non-government organizations, just like Japan and other Western countries do, to effectively help the poor in specific areas.

Notes

1. Flynt Leverett, "Resource Mercantilism and the Militarization of Resource Management", in *Energy Security and Global Politics*, edited by Daniel Moran and James A. Russell (Abingdon: Routledge, 2009).
2. Quoted from Paul Stevens et al., "Conflict and Coexistence in the Extractive Industries", *A Chatham House Report*, November 2013, <http://www.chathamhouse.org/publications/papers/view/195670> (accessed 26 March 2015).
3. Ahmad D. Habir, "Resource Nationalism and Constitutional Jihad", in *Southeast Asian Affairs 2013*, edited by Daljit Singh (Singapore: Institute of Southeast Asian Studies, 2013), p. 123.
4. David R. Mares, "Resource Nationalism and Energy Security in Latin America: Implications for Global Oil Supplies", James A. Baker III Institute for Public Policy Working Paper, Rice University, January 2010, <http://weber.ucsd.edu/~dmares/MaresResourceNationalismWorkPaper.pdf> (accessed 26 March 2015).
5. Ibid.
6. Ozawa Terutomo, *Multinationalism, Japanese Style: The Political Economy of Outward Dependency* (Princeton: Princeton University Press, 1979).
7. Lee Min Yong, "Securing Foreign Resource Supply: Resource Diplomacy of South Korea", *Pacific Focus* 3, no. 2 (1988): 79–102.
8. Quoted from Lin Chinxu and Shi Yinghong, "Hanguo haiwai ziyuang kaifa zhanlue" [South Korean Overseas Resource Development Strategy], *Journal of Contemporary Asia-Pacific Studies*, no. 2 (2014): 54–65.
9. Quoted from Ahmad D. Habir, "Resource Nationalism and Constitutional Jihad", in *Southeast Asian Affairs 2013*, edited by Daljit Singh (Singapore: Institute of Southeast Asian Studies, 2013), p. 123.
10. Ibid.
11. Lavina Lee, "Myanmar's Transition to Democracy: New Opportunities or Obstacles for India?", *Contemporary Southeast Asia* 36, no. 2 (2014): 290–316.
12. "Sino–Myanmar Crude Pipeline Memo Signed", *Downstream Today*, 19 June 2009, <http://www.downstreamtoday.com/news/article.aspx?a_id=16796&AspxAutoDetectCookieSupport=1> (accessed 26 March 2015).

13. "CNPC Announces Rights and Obligation Agreement Signed of Myanmar–China Crude Pipeline", *Your Project News*, 21 December 2009, <http://www.yourprojectnews.com/cnpc+announces+rights+and+obligation+agreement+signed+of+myanmar-china+crude+pipeline_43693.html> (accessed 26 March 2015).
14. Wu Lei, "Xioujian zhongmian shiyou guandao: ba Yunnan jianshe cheng woguo zhongyao de shiyou chucun jidi" [China–Myanmar Oil Pipeline Construction: Building Yunnan into China's Important Oil Storage Base], Research Report, Yunnan University, July 2005.
15. Author interview with Chinese scholars in Yunnan, August 2011.
16. Soe Than Lynn, "Myanmar Parliament Approves Letpadaung Mine Probe", *Myanmar Times*, 24 November 2012.
17. "Suu Kyi Speaks Out on Monywa Copper Mine Project", *Mizzima News*, 26 November 2011, <http://archive-2.mizzima.com/news/inside-burma/8451-suu-kyi-speaks-out-on-monywa-copper-mine-project.html> (accessed 26 March 2015).
18. Ibid.
19. Schearf Daniel, "Burma Recommends Controversial Mine Continue", *VOA News*, 12 March 2013, <http://www.voanews.com/content/burma-recommends-controversial-china-backed-mine-continue/1619836.html> (accessed 26 March 2015).
20. Tin Maung Maung Than, "Myanmar and China: A Special Relationship?", *Southeast Asian Affairs 2003* (Singapore: Institute of Southeast Asian Studies, 2003).
21. In late July 1956, more than a thousand Chinese troops entered the Wa State from Yunnan and remained entrenched following two bloody clashes with Myanmar troops along the disputed border in November 1955 and April 1956. U Nu's appeal to Chairman Mao during his visit to Beijing in October 1956 diffused the tense border situation and resulted in the withdrawal of the Chinese troops from Myanmar territory by the end of the year. Such an incident was not repeated as the border issue was settled within the next four years.
22. Tin Maung Maung Than, "Myanmar and China: A Special Relationship?", *Southeast Asian Affairs 2003* (Singapore: Institute of Southeast Asian Studies, 2003).
23. "Lu Qizhou: Central Government-owned Enterprises 'Going Global', Draw Lessons from the Suspension of Myitsone Dam", *Caixin.com*, 10 March 2012, <http://economy.caixin.com/2012-03-10/100366609.html> (accessed 26 March 2015).
24. Aung Tun, "Myanmar's 'Look West' Policy: Is China Being Sidelined?", *The Diplomat*, 26 June 2013.
25. Lynn Thiesmeyer, "Mekong River Transboundary Management Issues in the 21st Century: Yunnan Province and Its Southern Neighbors", paper presented

at the RSIS workshop on "China and Non-Traditional Security: Global Quest for Resources and Its International Implications", Singapore, 31 October 2014.

26. Elliot Brennan and Stefan Doring, "Myanmar's Dam May Be No Show", *Asia Times Online*, 4 March 2014.
27. Aung Shinb, "Environmentalists Call on Government to Halt Hydropower Projects", *Myanmar Times*, 3 November 2013.
28. Bernt Berger, "China's Troubled Myanmar Policy", *The Diplomat*, 23 August 2013.
29. Fan Hongwei, "Enmity in Myanmar Against China", *ISEAS Perspective*, 17 February 2014.
30. Ibid.
31. Naing Ko Ko and Simon Scott, "Rethinking Japan's Myanmar Policy", *Japan Times*, 7 July 2011.
32. Tin Maung Maung Than, "Myanmar in 2013: At the Halfway Mark", *Asian Survey* 54, no. 1 (2014).
33. "Japan and Southeast Asia: Hand in Hand", *The Economist*, 1 June 2013.
34. "Japan Searches for Business Possibilities in Myanmar", *China Daily*, 27 May 2013.
35. "Japan and Southeast Asia: Hand in Hand", *The Economist*, 1 June 2013.
36. IMF, *Directions of Trade Statistics Yearbook 2012.*
37. Ibid.
38. Ministry of Commerce of China, *2012 Statistical Bulletin of China's Outward FDI.*
39. Takamine Tsukasa, *Japan's Development Aid to China: The Long-Running Foreign Policy of Engagement* (Abingdon: Routledge, 2006).
40. Alan Rix, *Japan's Economic Aid: Policy-making and Politics* (New York: St. Martin's Press, 1980).
41. Mikio Oishi and Fumitaka Furuoka, "Can Japanese Aid be an Effective Tool of Influence? Case Studies of Cambodia and Burma", *Asian Survey* 14, no. 6 (2003).
42. Toshihiro Kudo, "China and Japan's Economic Relations with Myanmar: Strengthened vs Estranged", IDE-JETRO BRC Research Report, 2009, p. 274.
43. In the Japanese view, emerging donors, like China, are providing aid that is better directed to the needs of the recipient government while Western donors have concentrated their aid in the health and education sectors. Some emerging donors are not as concerned about the environment or human rights, so the long-term impact of their assistance ought to be prudently weighed. See Sato. Jin, "A Japanese Approach to Assistance: Cherishing the Recipient Experience", *Asahi Shimbun*, <http://www.asahi.com/shimbun/aan/english/hatsu/eng_hatsu111031.html> (accessed 30 June 2014).

44. Timur Dadabaev, "Chinese and Japanese Foreign Policies Towards Central Asia from a Comparative Perspective", *The Pacific Review* 27, no. 1 (2014).
45. Marie Lall, "Room for Both Japan and China to Do Business in Myanmar", *Global Times*, 21 January 2013.
46. *Japan's Official Development Aid White Paper 2012.*
47. Cited in James Reilly, "A Norm-Taker or a Norm-Maker? Chinese Aid in Southeast Asia", *Journal of Contemporary China* 21, no. 73 (2012): 71–91.
48. Ibid.
49. China's aid is financed by loans, while the United States and OECD subscribe to a definition of "development aid" that confines the term solely to grants.
50. *China's Official Development White Paper 2011.*
51. Thomas Lum, Hannah Fischer, Julissa Gomez-Granger, and Anne Leland, "China's Foreign Aid Activities in Africa, Latin America, and Southeast Asia", *Congressional Research Service*, 25 February 2009, <http://fas.org/sgp/crs/row/R40361.pdf> (accessed 30 June 2014).
52. Export-Import Bank of China, Annual Report 2006, <http://english.eximbank.gov.cn/annual/2012fm.shtml> (accessed 30 June 2014).
53. Xinhua News Agency, "FM: China to Remain Developing Country in Decade", <http://en.chinagate.cn/2012-03/06/content_24820263.htm> (accessed 1 March 2014).
54. Yang Jiechi, "Implementing the Chinese Dream", *The National Interest*, 10 September 2013, <http://nationalinterest.org/commentary/implementing-the-chinese-dream-9026> (accessed 30 June 2014).
55. John Lee, "Myanmar Pivots Awkwardly Away from China", *ISEAS Perspective*, 12 December 2013.
56. Mark Beeson, Mills Soko, and Wang Yong, "The New Resource Politics: Can Australia and South Africa Accommodate China?", *International Affairs* 87, no. 6 (2011).
57. Naing Ko Ko and Simon Scott, "Rethinking Japan's Myanmar Policy", *Japan Times*, 7 July 2011.
58. In the midst of the Cold War, the Ne Win government had long pursued a strict neutralist foreign policy, and refused to ally itself with any bloc.
59. Ma Yangbing, "Yinmian guanxi de fazhan ji dui zhongguo de yingxiang" [Indian–Myanmar Relations and the Impact on China], *Asia and Africa Review* (Beijing), no. 6 (2009).
60. Tuli Sinha, "Myanmar–China Energy Engagement: Implications for India", *IPCS Issue Brief* (New Delhi: Institute of Peace and Conflict Studies, December 2009).
61. *BP Statistical Review of World Energy*, June 2011.
62. "India Urges Burma to Release Opposition Leaders", *Reuters*, 4 August 1989.
63. Siddharth Varadarajan, "India Not Interested in Exporting Ideology", *Times of India*, 20 May 2007.

64. Gideon Lundholm, "Pipeline Politics: India and Myanmar", *PINR Power and Interest News Report*, 10 September 2007.
65. Lavina Lee, "Myanmar's Transition to Democracy: New Opportunities or Obstacles for India?", *Contemporary Southeast Asia* 36, no. 2 (2014): 290–316.
66. Ibid.
67. Tuli Sinha, "Myanmar–China Energy Engagement: Implications for India", *IPCS Issue Brief* (New Delhi: Institute of Peace and Conflict Studies, December 2009).
68. IMF, *Direction of Trade Statistics Yearbook 2012*.
69. Kong Bo, "The Geopolitics of Myanmar–China Oil and Gas Pipelines", *Pipeline Politics in Asia*, NBR Special Report #23, September 2010.
70. Chiung-Chiu Huang, "Balance of Relationship: The Essence of Myanmar's China Policy", *The Pacific Review* 28, no. 2 (2015): 189–210.
71. Ibid.
72. Jacob Goldberg, "Myanmar's Great Power Balancing Act", *The Diplomat*, 29 August 2014.
73. Quoted from Jacob Goldberg, "Myanmar's Great Power Balancing Act", *The Diplomat*, 29 August 2014.
74. Ibid.
75. McKinsey Global Institute, "Myanmar's Moment: Unique Opportunities, Major Challenges – Executive Summary", June 2013, <http://www.slideshare.net/naythiha/myanmars-moment-unique-opportunities-major-challenges-executive-summary> (accessed 26 March 2015).
76. "Extractive Industries Transparency Initiative", *Myanmar Times*, 7 August 2013.
77. Lavina Lee, "Myanmar's Transition to Democracy: New Opportunities or Obstacles for India?", *Contemporary Southeast Asia* 36, no. 2 (August 2014): 290–316.

5

CASE STUDY (2) Indonesia

Indonesia is another case example of China's energy relations with ASEAN countries. As the biggest country in Southeast Asia, Indonesia illustrates the diplomatic complexities in its relations with China. Although China has planned to stake a long-term strategic energy investment in the country, the fact that its major trading partner is also seem as a security uncertainty in the region has complicated the resource policies in Indonesia and encouraged domestic resource nationalism. Rising fears that the increasing unbalanced trade relations might affect its national economic security have stirred debates over how Indonesian mineral industries could remain competitive as the country continues its trade ties with Beijing. For the concern of being pulled into China's orbit in a dependent relationship based on supplying raw materials, Jakarta implemented a new law banning the export of unprocessed ore in January 2014. Although the aim is to increase the value added from mineral resources, the new regulations will certainly affect Sino–Indonesian energy cooperation.

OVERVIEW OF ENERGY SECTORS IN INDONESIA

Indonesia is rich in minerals and is in the top ten countries in the world for proven reserves of coal, copper, nickel, tin, bauxite, and coal. Indonesia produces more than 15 per cent of the global nickel supply and 3 per cent of the global copper supply, and it is the world's largest exporter of thermal coal. The total mineral export value more than tripled from

US$3 billion to US$11.2 billion in 2013, driven by historically high commodity prices and increasing production.[1] By value, approximately 40 per cent of total mineral export is currently processed; all tin exports are processed, while most copper, nickel, and bauxite exports are unprocessed.

Energy and mineral resources are important sectors in Indonesia's economy. The role of energy and mineral resources to the economy can be observed by three indicators such as its share to the gross domestic product (GDP) and its contribution to economic growth, export, and state revenue. The energy sector has become a buffer of national export and it has made significant contributions to the state revenue, both from tax and non-tax revenue. As seen from Table 5.1, the share of the three main energy sectors in GDP was about 11.5 per cent between 2010 and 2014, and crude oil, gas, and geothermal has the highest share. Between 2010 and 2014, the average economic growth was about 5.7 per cent and the coal sector contributed positively to economic growth while the two other sectors indicated a negative growth.

Table 5.2 indicates that the contribution of energy and mineral resources to the state revenue come from two main sources — tax and non-tax revenue. In 2015, the tax and non-tax revenue from oil decreased due to the declining of Indonesia crude oil price (ICP). However, the contribution from mineral and coal increased. As seen from Table 5.2, the contribution of energy to the state revenue has declined from about 33 per cent in 2006 to about 10 per cent in 2015. Certainly, this also indicates the rise of non-energy tax and non-tax revenue to the economy.

The role of the energy sector in supporting Indonesian economy has been changing due to the rapid increase in domestic energy consumption. Indonesia's oil reached peak of production in the early 1990s, then it gradually declined. The gas production reached a peak in 2010, then it declined. Indonesia is still the world's largest exporter of thermal coal although its domestic demand is increasing rapidly. Because the growth of gas and coal production is higher than consumption, in 2013, the share of export to production for liquefied natural gas (LNG) and coal was about 88 per cent and 73 per cent respectively.[2] Nevertheless, growing domestic energy demand has impelled the Indonesian government to secure energy production for domestic consumption, although for gas and coal, the production is still higher than consumption (see Table 5.3).

TABLE 5.1
Contribution of Energy Sectors to GDP
(Rp$ billion)

Energy Sector	2010	2011	2012	2013	2014	Average Annual Growth (%)*
Crude Petroleum, Natural Gas, and Geothermal	336,170	444,068	492,894	519,210	506,445	–2.4
Coal and Lignite Mining	160,733	253,026	270,519	275,988	251,303	11.3
Manufacture of Coal and Refined Petroleum Products	233,822	284,099	298,403	310,863	331,743	–1.6
GDP	6,864,133	7,831,726	8,615,705	9,524,737	10,542,694	5.7
Share of Energy Related Products to GDP (%)	10.6	12.5	12.3	11.6	10.3	

Note: At 2010 constant price
Source: Central Bank of Indonesia.

TABLE 5.2
Contribution of Energy and Mineral Resources to the State Revenue
(Rp$ billion)

Year	Tax Revenue from Oil and Gas	Non-Tax Revenue from Oil	Non-Tax Revenue from Gas	Non-Tax Revenue from Mineral and Coal	Non-Tax Revenue from Geothermal	Total State Revenue from Energy Related Sector	Domestic Revenue	Share of Energy Revenue to the Total State Revenue (%)
2006	43,188	125,145	32,941	6,781	—	208,055	636,153	32.7
2007	44,001	93,605	31,179	5,878	—	174,662	706,108	24.7
2008	77,019	169,022	42,595	9,511	941	299,089	979,305	30.5
2009	50,044	90,056	35,696	10,369	400	186,566	847,096	22.0
2010	58,873	111,815	40,918	12,647	344	224,597	992,249	22.6
2011	73,096	141,304	52,187	16,370	563	283,519	1,205,346	23.5
2012	83,461	144,717	61,106	15,877	739	305,901	1,332,323	23.0
2013	88,747	135,329	68,300	18,621	867	311,864	1,432,059	21.8
2014	83,890	154,750	56,918	23,560	580	319,697	1,633,053	19.6
2015	50,919	72,999	22,638	31,679	584	178,819	1,765,662	10.1

Source: Government state budget — various issues

TABLE 5.3

Growth of Oil, Gas, Coal Production and Consumption in Indonesia

(%)

	Oil		Gas		Coal	
	Production	**Consumption**	**Production**	**Consumption**	**Production**	**Consumption**
Year	**1991–2013**	**1965–2013**	**1970–2013**		**1981–2013**	
Growth at corresponding year	–3.2	5.8	9.4	8.2	20.7	16.3

Source: Calculated from *BP Statistical Review of World Energy June 2014*.

POLITICAL AND ECONOMIC COOPERATION

Indonesia was the first Southeast Asian country that established official diplomatic relations with China in July 1950. The first fifteen years of that relationship, however, were replete with problems and suspicion, which culminated in Indonesia's decision to freeze diplomatic ties in October 1967. The resumption of diplomatic relationship in August 1990 did not immediately remove the thorny issues between the two countries. Suspicion and sensitivity continued to characterize Indonesia's attitude toward China. Although Indonesians began to recognize the importance of China, the problem of their country's ethnic Chinese minority continued to affect its perceptions of China. Indonesia still worried about the possible link between the People's Republic of China and the ethnic Chinese minority. The perceptions of the general public and political elite also continued to be coloured by various stereotypes associated with the minority. Worse, there was a new dimension in Indonesia's view: worry about China's regional role and policies in Southeast Asia, especially regarding Beijing's behaviour in the South China Sea and China's growing military capability.[3]

It was only after 1998, however, that Indonesia-China relations began to show significant signs of improvement and closer cooperation due to dramatic changes in Indonesia's domestic politics since May 1998 and China's "good neighbour" policy and "charm diplomacy" toward Southeast Asia (including Indonesia) — as demonstrated in the aftermath of the 1997 financial crisis and during the 2004 tsunami. A newly democratizing Indonesia seemed to pursue a very different attitude and policy toward China. Since President Abdurrahman Wahid, who became the country's first democratically-elected president in October 1999, the imperative for improving relations had become a matter of importance in the foreign policy discourse of every successive government in Jakarta. The basis for active bilateral cooperation received a stronger impetus when, on 25 April 2005, Jakarta and Beijing signed an agreement to establish a strategic partnership agreement[4] to foster socio-economic and military cooperation. During President Susilo Bambang Yudhoyono's visit to China in July of the same year, officials concluded several major agreements covering not only traditional areas of cooperation in trade and investment but also defence technology cooperation, suggesting that China–Indonesia relations have come full circle.

Indeed, at the bilateral level, Indonesia has increasingly become more comfortable and confident in dealing with China. It now no long sees

China as a threat to Indonesia's national security and internal stability. Instead, China has become a reference for success. More and more Indonesians see China, compared with the United States, as an increasingly positive cooperator. For example, the Pew Research Global Attitudes Survey released in 2013 showed that the China's favourability in the eyes of Indonesian respondents increased from 58 per cent in 2010 to 67 per cent in 2013, while the number of respondents with a favourable view of the United States rose slightly from 59 per cent to 61 per cent; and 69 per cent of the respondents replied that China will have a great deal of influence on the way things are going in Indonesia, increasing from 60 per cent in 2008.[5]

Moreover, the survey also showed that 54 per cent of the respondents agreed that China has considered Indonesia's interest when making international policy decisions, increasing from 50 per cent in 2008. In comparison, only 52 per cent indicated that the United States takes into account Indonesia's interest in making international policy decisions. Although Indonesian elites like the idea of U.S. engagement in the region and dislike the thought of a dominant Chinese role, they have far more confidence in the Chinese commitment to the region that they do than in the U.S. commitment.[6] Most Indonesians no longer see China as an ideologically threatening state, but as an economic opportunity and challenge. China's growing economy fits well with Indonesia's current requirements, as President Yudhoyono has stated, that "our target in developing relations with China is to look for an opportunity to fulfill our national interests. We have to get something from the rise of China, especially in economic terms."[7]

China's perceptions of Indonesia have also changed significantly. "Beijing clearly appreciates the fact that new governments in Jakarta no longer need to cultivate the China threat thesis, practiced for more than two decades by Suharto's regime, as the basis of regime legitimacy."[8] In Chinese eyes, Indonesia's image as an anti-Chinese nation has begun to fade away, even though some residual negative perceptions remain because of the anti-Chinese riots in 1998. Today, China sees Indonesia as a critical country in Southeast Asia; a close relationship, which would greatly benefit China's economic, political, and strategic interests in the region, is an integral part of Beijing's engagement policy with Asia.[9]

The evolution of mutually positive perceptions laid a good foundation for economic cooperation between these two countries. The growing bilateral economic engagement can be gauged from the fact that despite the global meltdown, the two countries achieved the target of bilateral trade of US$30 billion in 2008. The bilateral trade increased from US$19 billion in 2006 to US$68 billion in 2013, registering a more than 300 per cent growth in the seven years. The countries have agreed to increase the volume of bilateral trade to US$80 billion by 2015.[10] Increasing bilateral trade has helped Indonesia reduce its over-dependence on Western markets. Due to the expanding trade with China, Indonesia's over-reliance on particular export destination countries has decreased. For example, from 2000–12, the market shares of the United States, Japan, and Europe had decreased from 51 per cent to 37 per cent, while China's share increased from 3.6 per cent to 12 per cent.[11] It was Asian emerging economies, mainly China, India, and ASEAN, that had subsequently compensated Indonesia's decelerating exports to developed countries.

SINO–INDONESIAN ENERGY RESOURCE TIES

Energy Cooperation

The fast-growing China–Indonesia political and economic relations largely reflects two trends — China's growing demand for energy and mineral resources and consequent China's increasing investment in Indonesia. Although Indonesia is not the only ASEAN country that boasts natural resources that China needs, it is a vital focal point of Beijing's resource diplomacy.

Bilateral energy cooperation between China and Indonesia is not new. It can be traced back to the 1980s when China's NOCs (National Oil Companies) developed partnership through merge and acquisition with other NOCs and IOCs (International Oil Companies). In February 1988, PetroChina signed the offshore production sharing contract (PSC) with Indonesia. The contract area was located in Tuban, East Java. In 1994, China National Offshore Oil Corporation (CNOOC) obtained 2.8 per cent share of Indonesian Malacca oilfield, Offshore Indonesia, through capital mergers and acquisitions, starting its first entry into Indonesian energy exploration and development markets; in 2002 CNOOC bought a Spanish oil company's assets in Indonesian oil fields at a price of US$850 million, thus becoming Indonesia's largest offshore oil producer; in April 2004, Sinopec (China Petroleum and Chemical Corporation) purchased American Devon Energy's oil and gas assets in Indonesia as its first entry into Indonesian energy exploration and development market.[12] In 2005, CNOOC obtained 16.9 per cent of the shares of a British Gas Corporation LNG project in Indonesia.[13] The process of acquisition continued. In 2008, CNOOC purchased interest in Husky's Indonesia project — owned by Canada's largest energy company — in order to explore the deep water blocks. In addition, the Sinopec and other Chinese NOCs are also interested in constructing storage facilities in Indonesia. The total acquisition in Indonesia between 2002 and 2011 was about US$2.45 billion, or about 1.6 per cent of the total Chinese overseas oil and gas upstream acquisitions.[14]

Although the share of Chinese overseas oil and gas upstream acquisitions in Indonesia and the inflow of investment from China were minor, China's investment flow to Indonesia's mining sector has been increasing rapidly. The latest figures below indicate that most of Chinese FDI (foreign direct investment) has flowed to the mining sector. According to Table 5.4, in 2014, the share of China's FDI to the mining sector accounted for 99 per cent of its total FDI to Indonesia (Share 1). In the total FDI flows to the mining sector, China's share had increased from 20 per cent in 2005 to 39 per cent in 2014 (Share 2), although the share of its FDI had only increased from 3.6 per cent to 4.8 per cent (Share 3).

The reason for the increase of China's FDI in the mining sector is mainly because of China's increased demand for coal. When China became a net importer of coal in 2007, it shifted its focus to Indonesia thereafter. Coal from Indonesia became increasingly attractive to the prosperous coastal regions of China, potentially displacing domestic Chinese production that must be railed and shipped long distances from Shanxi and Mongolia. As a part of the growing effort by Chinese companies to secure future coal supply, in July 2010, Shenhua — China's largest coal producer — announced a US$331 million coal project in Sumatra, and in October the same year, China's sovereign wealth fund injected US$1.9 billion into Bumi Resources, Indonesia's largest coal producer.[15]

Energy cooperation between China and Indonesia through equity capital and investment, has brought more energy trade between Indonesia and China. As Table 5.5 shows, the total export of energy related products (HS-27) between 2006 and 2011 increased from US$27.6 billion to US$69 billion. We can see that as the total mineral export increased, the shares of mineral export to Japan and Korea declined. While the share of Indonesia's export to China was relatively stable, and the value increased steadily from US$3.1 billion in 2006 to US$8.3 billion in 2013, or on average it increased by 24 per cent per year. This indicates that there have been changes in Indonesia energy's export market, with China and India becoming more important markets.

TABLE 5.4
FDI Inflows to Indonesian Mining Sector
(US$ million)

Year	Total FDI Inflows from China	FDI Inflows from China in Mining Sector	Total FDI Inflow	Total FDI Inflows in Mining Sector	Share 1	Share 2	Share 3
	(1)	(2)	(3)	(4)	(2) : (1)	(2) : (4)	(1) : (3)
2005	299.5	238.8	8,336.0	1,225.8	79.7	19.5	3.6
2006	124.0	123.0	4,914.0	322.1	99.2	38.2	2.5
2007	178.0	170.0	6,928.0	1,904.0	95.5	8.9	2.6
2008	531.0	534.0	9,318.0	3,609.5	100.6	14.8	5.7
2009	359.0	357.4	4,877.0	1,301.1	99.6	27.5	7.4
2010	354.0	354.0	13,771.0	1,896.0	100.0	18.7	2.6
2011	215.0	150.0	19,242.0	3,418.0	69.8	4.4	1.1
2012	335.0	285.0	19,138.0	1,822.0	85.1	15.6	1.8
2013	67.6	32.2	18,947.8	2,486.8	47.7	1.3	0.4
2014	1,066.9	1,065.9	22,276.3	2,710.7	99.9	39.3	4.8

Note: Share is in percentage.
Source: Calculated from the Bank Central Republic Indonesia.

TABLE 5.5

Share of Mineral Fuels, Oils and Products of Their Distillation (HS:27): Export from Indonesia to Major Countries

Year	China (%)	India (%)	Japan (%)	Republic of Korea (%)	Singapore (%)	Total 5 Countries (%)	Total Export HS 27 (billion US$)
2006	11.2 (3.1)	2.5	39.4	18.2	4.5	75.8	27.6
2007	12.0 (3.5)	3.0	40.6	16.6	5.9	78.1	29.2
2008	11.5 (4.6)	3.4	40.3	14.6	7.0	76.8	39.8
2009	14.1 (4.7)	6.2	26.7	14.8	7.2	68.9	33.0
2010	12.9 (6.0)	5.3	25.8	17.9	9.0	70.9	46.8
2011	12.9 (8.9)	6.9	27.8	16.9	10.6	75.2	68.9
2012	12.8 (8.1)	7.9	26.0	17.3	10.4	74.5	63.4
2013	14.4 (8.3)	9.7	24.8	13.1	11.0	72.9	57.4

Note: Figure in brackets indicates in billion US$.
Source: Calculated from the United Nations Commodity Trade Statistics Database.

Resource Nationalism versus Domestic Politics

In Indonesia, which was considered by many Chinese investors to be one of the most promising investment destinations, recent years found that there have been an increase in regulation changes in a variety of sectors ranging from mining to oil. As such, Indonesia is increasingly described as a country where resource nationalism is on the rise.[16] Law 4/2009 on Mineral and Coal Mining (the 2009 Mining Law) is the most widely referenced example of Indonesia's rising resource nationalism. The 2009 Mining Law and its implementing regulations replaced the decades old Contract of Work (CoW) system. The CoW system was widely viewed as offering favourable and stable conditions for foreign investors since it allowed foreigners to obtain full mining licences. In contrast, the new law requires foreign companies to reduce their ownership in mines to a maximum 49 per cent by the tenth year of production, thus effectively stripping foreigners of their control over mining assets.[17] It has sought to reduce the share of foreign ownership in Indonesian mining assets, forcing foreign investors with mining business licences and special mining business licences to divest their stakes in Indonesian mines.

In addition to the divestment policy, the government has also placed stronger restrictions on the export of raw materials. The Indonesian government argued that by imposing export duties, government can control trading on raw material or ore, increase value added, and ensure availability of mineral resources for domestic market.[18] It has imposed export duties to ensure that it collects some revenue from mineral extraction, and it has stipulated that raw materials must be processed domestically from 2014.[19] It also places a ban on the export of certain raw minerals and requires mining companies to build smelters for domestic processing. The new regulations require reserve coal for the domestic market, stipulating that coal miners will have to upgrade the heating value of their coal if they want to continue to export.[20] The export ban was finally implemented in January 2014, causing significant impact debates in Indonesia.

As seen from Table 5.6, the total export of minerals declined significantly after the government implemented the export duties in 2012. The export of seven mineral products reached a peak of US$44.5 billion in 2011, then it declined substantially to US$29.6 billion in 2014. The share of minerals in total exports also decreased from about 27 per cent to 20 per cent between 2010 and 2014.

TABLE 5.6
Exports of Selected Mineral Products
(US$ million)

No.		2010	2011	2012	2013	2014
1.	Mineral fuels & oils	18,726	27,444	26,408	24,780	21,058
2.	Cooper	3,306	3,811	1,886	1,738	1,967
3.	Ores, Slag, and Ash	8,149	7,343	5,083	6,544	1,919
4.	Tin	1,735	2,439	2,132	2,129	1,814
5.	Iron and steel	1,102	1,353	875	652	1,148
6.	Nickel	1,436	1,218	993	942	1,058
7.	Aluminium	772	869	784	693	665
	Total export of 7 commodities (Nos. 1–7)	35,224	44,476	38,161	37,479	29,628
	Share in total exports	27.15	27.45	24.93	25.00	20.30

Source: Calculated from *Economic Profile*, Ministry of Trade, <http://www.kemendag.go.id/en/economic-profile/indonesia-export-import/growth-of-non-oil-and-gas-export-commodity> (accessed 16 March 2015).

In reality, Indonesia is in a dilemma on this export ban policy. The Indonesian government recognizes the importance of shifting from an economic model heavily reliant on raw material exports to one in which Indonesia refines its own metals, minerals, and ores, both for exports and, more importantly, for domestic use in manufacturing industries. The underlying economic rational put forward by the government for the ban is to stimulate the domestic smelting and processing capacity, which will lead to significantly higher value addition in mineral exports.[21] But such interventions come with substantial risks: the industry may respond to the incentives differently from the policy intention. Foreign investors may reduce their investment, as investors in Indonesia (both domestic and foreign) have often chosen to export raw commodities, because other countries already have well-developed processing capabilities. Most foreign investors do not support the new divestment rules. Investment in the mineral sector is often a long-term proposition, so companies may not want to be involved in a project over which they will have little control in the future.[22]

Without adequate investment and capital inflows, Indonesia finds itself squeezed between slowing foreign demand for raw materials and the inability to readily shift to greater domestic consumption of key minerals. Overseas processors are less compelled to invest in Indonesia as they can secure ore supplies elsewhere (Indonesia accounted for less than 2 per cent of global production in copper, lead, and zinc in 2012 and does not have a major share of reserves for any of these commodities[23]). Investment in bauxite and iron ore are more likely to be viable if the raw ore can be accessed cheaply, placing Indonesia at a disadvantage compared to other countries such as China and India. Thus "the export ban will inevitably lead to a dramatic decline of output in Indonesia's extractive industries, damaging foreign investment and economic growth, and disrupting global mineral markets".[24]

The new regulations, especially the export ban of raw materials, will certainly affect Sino–Indonesian energy resource cooperation. China is highly dependent on Indonesia for its nickel, bauxite, copper, and coal. In 2012, Indonesia produced 16 per cent of the world's nickel ore, and supplied 58 per cent of the world's nickel import demand, and 48 per cent of bauxite import demand. Most of Indonesia's exports of these metals go to China and Japan. For China, in 2013, China sourced 66 per cent of its aluminium ore (bauxite is refined into alumina before being turned into aluminium) from Indonesia (up from 64 per cent in 2012) and 57 per cent of its bauxite ore (on par with 2012 level). In 2012, about 6 per cent of China's copper ore imports came from Indonesia.[25]

The natural resource sectors have largely steered Indonesia's economic growth over the past decade, with energy related products contributing roughly 10 per cent of Indonesia's GDP and being a critical source of employment and tax revenue (US$9.1 billion in 2011, with coal alone providing US$7.7 billion, nearly 10 per cent of total tax revenue). As seen from Table 5.7, the share of energy related products to the total export was about 25 per cent in 2014. Currently, coal has become the major source of export revenue from energy sector.

Indonesia is still one of the largest exporters of coal. In 2012, its coal production was 237 million tons of oil equivalent (Mtoe), and its consumption was 50.4 Mtoe, increasing by 15 per cent per year on average since 2000.[26] The sharp increase in Indonesia's coal output has been driven by surging demand in the international market. Since 2000, Indonesia has accounted for almost 60 per cent of the growth in the world coal trade, and it became the largest exporter of steam coal in 2000.[27] In 2011, it overtook Australia as the world's largest exporter of coal. Indonesia plays an important role in world coal markets, particularly as a regional supplier to Asian markets. Nearly all of its exports go to Asia, the major destinations in 2011 being China (31 per cent), India (22 per cent), Korea (21 per cent), Japan (11 per cent), and Taiwan (8 per cent).[28] China and India are two of Indonesia's largest export markets, accounting for 31 per cent and 22 per cent of Indonesia's total coal export in 2011 respectively.[29] China is driving Indonesian coal export growth, but the appetite for Indonesian coal in China is gradually reducing, as in China, the government has discussed a ban on certain coal imports with low energy content, while favouring higher quality Australian and South African coal.[30] In this sense, the export ban in Indonesia might have less impact on China but more impact on Indonesia itself.

Many observers believe that Indonesia's resource nationalism is in fact driven by political motivations, reflecting an ongoing struggle between central, provincial and local governments for control over the issuing of mining permits.[31] The struggle dates back to 1998, when Soeharto's fall thrust Indonesia toward decentralization. Thirty-one years of highly centralized governance based in Jakarta under Soeharto had led to deep social and economic imbalances between Java and the outer islands.

In response to calls for political and economic decentralization, in 1999, President Bacharuddin Jusfu Habibie moved to limit the central government's authority to matters of military and policy security. Provincial governments were granted limited independence from Jakarta on social

TABLE 5.7
Contribution of Energy Sector to Export
(US$ million)

	2005	2006	2007	2008	2009	2010	2011	2012	2013	2014
Coal	4,179	6,190	6,977	10,305	13,765	17,801	26,924	26,248	24,359	20,814
Crude oil	7,259	7,911	9,380	11,442	8,008	11,219	14,166	12,723	12,188	8,840
Natural gas	10,243	11,863	12,165	16,254	9,778	12,968	18,196	17,671	15,689	14,942
Liquefied natural gas	8,734	9,953	9,722	12,785	7,188	9,432	12,961	11,943	10,568	10,294
Total export	86,995	103,528	118,014	139,607	119,645	158,074	200,787	188,496	182,089	175,290
Share of energy related products to total export (%)	24.9	25.1	24.2	27.2	26.4	26.6	29.5	30.0	28.7	25.4

Source: Central Bank of Indonesia.

policies, while local and district governments gained control over economic policies, including control over the issuing mining permits. As in provinces such as East Kalimantan and Southeast Sulawesi, mining plays a big role in the local economy, conflicting claims over the control of extractive projects became a source of political tension between the central and regional governments. In 2011, President Yudhoyono released the country's "Master Plan" for economic development through 2025, calling Indonesia to transform itself from a natural resources exporter to an industrial manufacturing hub. In pursuing this goal, a degree of recentralization is needed to facilitate the long-term strategic planning.

Some observers also believe that resource nationalism in Indonesia can be leveraged for the officials' electoral ambitions. Deliberations over the draft of the 2009 Mining Law occurred in the lead-up to Indonesia's 2009 Presidential Election. The same argument was being put forth as campaigns heated up for the July 2014 Presidential Election and Indonesia announced an export ban on raw ore in January 2014. Clearly, the logic is that a strong nationalist agenda that privileges domestic industry over foreign investors appeals to populist sentiment and can win more votes, particularly when it concerns the ownership of natural resources.

Concerns over Cooperation with China

The development of Indonesia-China energy cooperation is far from being smooth. Public debate on the Indonesia-China energy trading agreement arose in 2009. One of the debates was focused on the LNG price. Most Indonesians argued that the price agreement on shipping the LNG from Tangguh to Fujian, China was far below the market price. For example, in 2002, the price agreement was US$2.4/MMBtu (million British thermal units) and maximum ceiling price for oil was about US$25/barrel. In 2006, the market price of LNG increased to US$3.35/MMBtu and the oil price was US$38/barrel. In 2006, the first renegotiation was conducted by the government, and the LNG price increased to about US$3.3/MMBtu. The second renegotiation was made in 2010, but failed. After Indonesian President Yudhoyono met the Chinese President Hu Jintao in 2012, the former asked the Minister of Energy and Mineral Resources to renegotiate the contract with CNOOC. Finally, in June 2014, the negotiation was concluded and Indonesia obtained a better price on LNG that was about US$8.65/MMBtu.[32]

TABLE 5.8
China's Trade Volume with ASEAN-5
(US$100 million)

	2005	2006	2007	2008	2009	2010	2011	2012	2013
Malaysia	307	371	465	535	518.6	742	899	948	1061
Singapore	333	409	472	524	478.3	571	631	693	759
Thailand	218	277	346	412	381.7	530	647	697	713
Indonesia	168	191	250	316	282.8	428	606	662	683
Philippines	176	234	306	286	205.3	278	323	364	381

Source: IMF, *Direction of International Trade Statistics Yearbook 2012*; China Customs Statistics Monthly.

The other concern is an increasing trade deficit with China. As seen from Table 5.8, Indonesian trade with China was slightly higher than Vietnam, but less than Singapore, Malaysia, and Thailand. Among the ASEAN countries, Indonesia had the highest trade deficit with China after Vietnam. Jakarta believes that growing non-minerals trade deficit with China is the main impediment to the expansion of bilateral trade. According to Indonesia's data, it had US$820 million trade surplus with China in 2005, but US$14 billion deficit in 2014 when oil and gas sectors were excluded.[33]

The bilateral trade structures should largely be blamed for Indonesia's rising trade deficit. Figure 5.1 shows that most of Indonesia's exports to China are resource-intensive products. For example, Indonesia's mining products exported to China as a proportion of its total exports to China had increased from 26.2 per cent in 2000 to 56 per cent in 2013. While in its total imports from China, high value-added products such as electrical machinery and transport equipment accounted for over 50 per cent (see Figure 5.2). In this sense, it is understandable that resource nationalism in Indonesia is easily on the rise.

Certainly, this increasing trade deficit should also be largely attributed to the low competitiveness of Indonesia's manufactured products. There is little evidence that the Indonesian government assisted firms in upgrading their technological capabilities in the 1980s and 1990s. By the mid-1990s, Indonesia lagged behind its East Asian neighbours on most technology indicators. According to Indonesian statistics, its spending

FIGURE 5.1
Commodity Structures of Indonesia's Exports to China, 2013

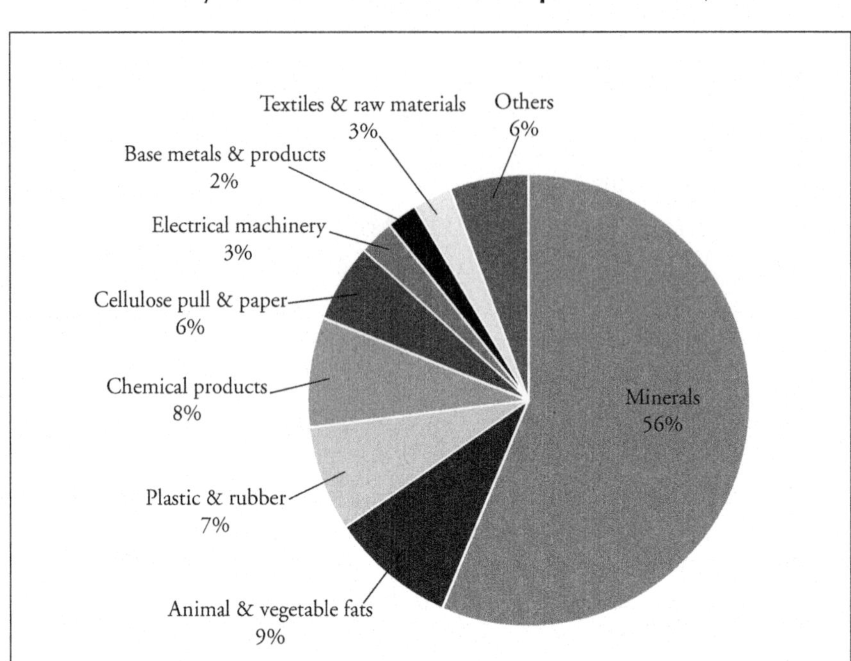

Source: Ministry of Commerce of China, *Country Report*, 2014.

on research and development was very low (0.2 per cent of GDP); it had very few patent applications (12 between 1981 and 1990); very few scientists and engineers were engaged in research and development (183 per million of the population); enrolments in tertiary education were low (10 per cent of the relevant age group in 1991); and few young adults had science or engineering degrees (0.4 per cent of 20–23 year olds).[34] Technological catch-up also requires exposing domestic enterprises to the rigours of foreign competition. Because of this requirement, policymakers in China unilaterally liberalized its trade and investment regime as much as, if not more than, their Indonesian counterparts did.[35] But rising fears that this widening trade gap might affect its national economic security have stirred debate over how Indonesian industries could remain competitive as the country seeks improved trade ties with Beijing, and in turn, this has aroused domestic economic and resource nationalism.

FIGURE 5.2
Commodity Structures of China's Exports to Indonesia, 2013

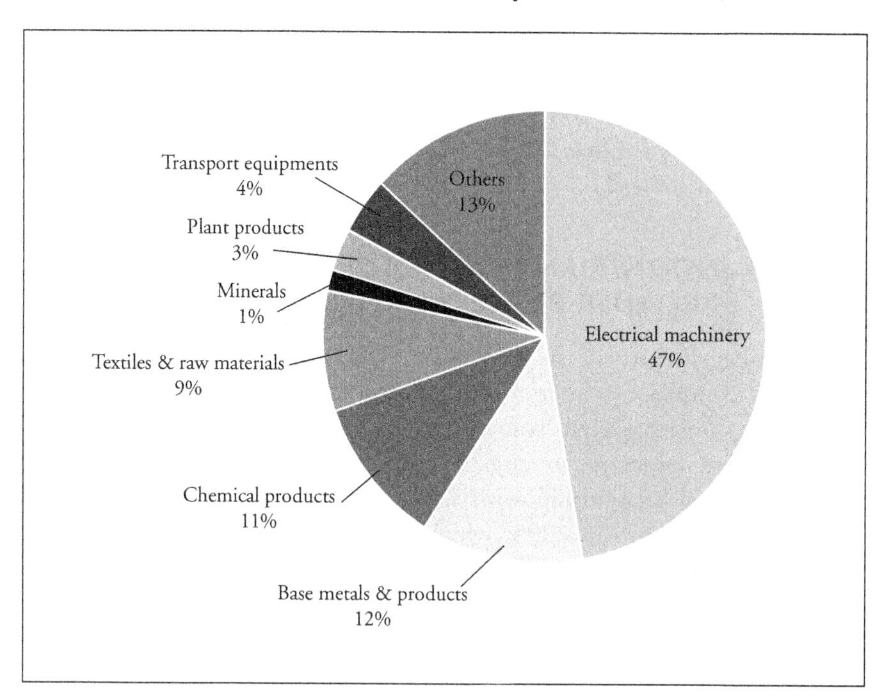

Source: Ministry of Commerce of China, *Country Report*, 2014.

One more concern is China's investment. China's overall investment commitment in Indonesia remains comparatively small although it expands quickly after 2010. For example, the value of its total outward FDI reached US$68.8 billion in 2013, the value of its FDI in ASEAN reached US$7.27 billion, a 19.1 per cent increase over the previous year, while the value of its FDI in Indonesia was US$1.6 billion.[36] In 2011, Japan's FDI in Indonesia reached US$5.2 billion, Australia's reached US$4.2 billion.[37] Part of the reasons for the slow Chinese capital inflow is that as the bilateral economic ties grow, there remains equally growing concerns. There are concerns and worries about the nature of Chinese FDI in Indonesia. For example, many Indonesians asked: "Chinese technology is cheap, but is it environmentally friendly?"; "Many projects also involve the use of the Chinese workforce, so there is a sovereign issue because you are not giving Indonesian labor a chance."[38] Indonesians would expect to see further upstream investment in Indonesian facilities by Chinese producers as they attempt to hedge the ongoing potential risk of full Indonesian export ban on unprocessed materials. But "the current situation is that many Chinese companies have rushed into mass action, with a lack of a prudent attitude and due diligence"; "many Chinese enterprises are only optimistic about investment in mineral revenue, but are not concerned about the risks".[39] These concerns are the main factors inhibiting a decision on the part of the Indonesian government to further relax restrictions on Chinese capital inflows.

SINO–INDONESIAN ENERGY TIES EXTENDING TO A BROADER BILATERAL RELATIONSHIP?

As the biggest country in Southeast Asia, Indonesia illustrates the diplomatic complexities in its relations with China. At the bilateral level, Indonesia has increasingly become more comfortable with China. Although initially being reluctant to engage with China, Indonesia has forged a closer bilateral relationship with China, culminating in the signing of a strategic partnership in 2005, which was upgraded to a comprehensive strategic partnership during Chinese President Xi Jinping's visit to Jakarta in 2013, and has also encouraged Beijing's close relations with ASEAN. Based on common understanding on regional economic integration in East Asia, the partnership aims at consolidating not only bilateral relations, but also an ASEAN-led cooperative community-building process. While ASEAN occupies centrestage in Indonesia's foreign policy, China supports the ASEAN-led multilateralism in East Asia. Beijing's acceptance of the centrality and significance of ASEAN has helped Indonesia to see China as a "natural partner" in the region.[40] The two countries have come together to strengthen ASEAN-driven cooperative processes in East Asia, help moderate hostile tendencies in Asia, and represent continental Asia in global multilateral initiatives. Beijing understands that strengthening relations with Indonesia will surely help to forge closer ties between China and ASEAN as a whole, and benefit China in achieving some policy objectives and strategic interests more successfully.

While the South China Sea issue is important to Indonesia, Jakarta would still strongly prefer not to let this single issue get in the way of burgeoning Sino–Indonesian ties. While welcoming the U.S. rebalancing toward Asia, some in Indonesia have raised concerns that Washington has placed too much emphasis on the military dimension of this strategy. From

Jakarta's perspective, the importance Washington attaches to Indonesia and ASEAN should not simply be derivative of China's rise but instead be based on the intrinsic value of the country and the subregion.[41] As the largest country in Southeast Asia and the world's largest Muslim-majority nation, Indonesia has sought to enhance its international footprint, particularly in key bilateral relationships, and would like to act as an intermediary between the parties involved in the territorial disputes in the South China Sea.

Based on mutual need and common interests, the relationship between Indonesia and China is likely to become stronger and grow further in the future. Viewed through China's lens, Indonesia's bountiful mineral wealth has elevated the relations between Jakarta and Beijing to a position of strategic importance. Moreover, in Beijing's view, recent political reform and economic growth has made Indonesia re-emerge on both the international and regional stage with expanded prestige both in the East and West. As it bolsters its strength, Indonesia's weight and importance in the region's balance of power will only grow, particularly with respect to China and the United States.

In Indonesia's strategic calculations, China's importance lies primarily in its being a growing source of foreign investment that Indonesia desperately needs to develop domestic natural resources and infrastructures. Indonesia needs a huge amount of investment in its energy sectors, including energy-related infrastructures like gas pipelines and sea ports. According to the Indonesia Medium Term Development Plan (2015–19), the Indonesian government has three top priority sectors to develop, these include food, energy, and maritime resources. On the energy structure, oil and coal production will decline, while natural gas will become the future of primary energy supply for Indonesia. In response to these targets, government plans to develop gas infrastructures such as pipe lines, gas stations, and the city gas network. Connecting supply locus and market locus among the islands is one of great challenges in optimizing gas utilization. Most of the gas is produced in the eastern part of Indonesia, and it needs to be shipped by sea to the western part. But Indonesia's poor infrastructure has been the major problem and challenge.

For example, while its overall index has improved over the past years, the country's infrastructure index remains very low: 76th for physical infrastructure; 103rd in terms of ports quality; and 98th in electricity supply.[42] World Bank study (2010) found that the cost of shipping a 40-foot container from Padang to Jakarta is US$600 while the same container can be shipped from Jakarta to Singapore (three times the distance between Padang and Jakarta) for only US$185.[43] The quality of port facilities remains alarming and shows no sign of progress, and the electricity supply continues to be unreliable and scarce. China with its total outward FDI being US$101 billion in 2013 has a big potential role to play in Indonesia's infrastructure sectors.

CONCLUSION

Although the energy relations between China and Indonesia have thus far generally proved to be mutually beneficial, concerns and uneasiness among Indonesians about the nature and impact of the relations are still prevailing. Like the case in Myanmar, the extent to which Indonesia will continue to satisfy China's resource hunger will depend on whether domestic societies in Indonesia feel that their concerns about China's influence on Indonesia's political economy are being addressed. Particular areas of concerns are the continuing impact of Chinese FDI on local jobs and environment, the erosion of the competitive advantage of Indonesian companies (in local and regional markets) by the growing presence of Chinese companies, the unbalanced trade relations, and perceptions that expanding commercial relations have exerted baleful influence on Indonesian foreign policy.

But compared with Myanmar, the dynamics in the overall relations between Indonesia and China are rather different. Myanmar is constrained by other priority considerations such as how to successfully conclude lasting peace settlements with armed ethnic minority groups. Myanmar's historical issues, particularly its efforts at national reconciliation with dozens of ethnic minority groups, directly and indirectly affect investors. In addition, as Myanmar opens up, the entry or resurgence of more global players like Japan and India provides Myanmar with more options, facilitating its attempts at loosing itself from the Chinese grip. What is different in the Indonesian case is that there are many other positive factors that may make bilateral resource politics more productive. China and Indonesia have upgraded their bilateral relationship from "strategic partnership" to "comprehensive strategic partnership". Both countries are keen to assert themselves on the international and regional stage, and can position themselves as part of a new world order that is more representative of contemporary geopolitical realities. The strategic potential of China's investment in the energy related-infrastructures and sea ports is not limited to enlarging Sino–Indonesian energy trade, but extends to Indonesia–China relations more broadly and fits Indonesia's ambition as a maritime power.

Notes

For this chapter, the author would like to thank Dr Maxensius Tri Sambodo for providing many important figures and views on Indonesian energy development.

1. The World Bank, "Investment in Flux", *Indonesia Economic Quarterly*, March 2014, p. 19.
2. It is calculated from the Indonesian Energy Handbook, Ministry of Energy and Mineral Resources.
3. Rizal Sukma, "Indonesia–China Relations: The Politics of Re-Engagement", *Asian Survey*, August 2009.
4. This was upgraded to a Comprehensive Strategic Partnership after Xi Jinping's visit in 2013.
5. Global Indicators Database, <http://www.pewglobal.org/database/indicator/24/country/101/> (accessed 26 March 2015).
6. Bates Gill, Michael Green, Kiyoto Tsuji, and William Watts, *Strategic Views on Asian Regionalism: Survey Results and Analysis* (Washington, D.C.: CSIS, February 2009), p. 15.
7. Quoted in Rizal Sukma, "Indonesia–China Relations: The Politics of Re-Engagement", *Asian Survey*, August 2009.
8. Rizal Sukma, "Indonesia–China Relations: The Politics of Re-Engagement", *Asian Survey*, August 2009.
9. Tang Shiping, "Grand Strategy: Searching for China's Ideal Security Environment", *Zhanlue yu Guanli* [Strategy and Management], no. 6 (December 2000).
10. Zhou Yan, "Indonesia Seeks More Chinese Investment", *China Daily*, 3 May 2011.
11. Based on IMF, *Direction of Trade Statistics Yearbook 2012*.
12. Li Tao, "Qian xi zhongguo-dongmen de nengyuan hezuo" [An Analysis of China–ASEAN Energy Cooperation], *Southeast Asian Studies*, no. 3 (2006).
13. Zhao Ping, "Shiyou jingkou zhanglue da tishu" [Speeding Up Oil Strategy], *Chinese Foreign Investment*, no. 8 (2005).
14. Julie Jiang and Chen Ding, *Update on Overseas Investments by China's National Oil Companies: Achievement and Challenges since 2011* (Paris: International Energy Agency, 2014).
15. Anthony Deutsch, "Asia Giants' Scrabble for Coal Reaches Indonesia", *Financial Times*, 9 September 2010.
16. Eve Warburton, "In Whose Interest? Debating Resource Nationalism in Indonesia", *Kyoto Review of Southeast Asia*, Issue 15 (March 2014).
17. Mateo Cabello, "Indonesia: Mining White Paper", Oxford Policy Management, November 2013.
18. Ministry of Finance Regulation, <http://www.jdih.kemenkeu.go.id/fullText/2012/75-PMK.011-2012Per.htm> (accessed 17 March 2015).

19. P.J Burke and B.P. Resosudarme, "Survey of Recent Developments", *Bulletin of Indonesian Economic Studies* 48, no. 3 (2012).
20. Coal upgrading is simply the removal of water content from low-calorific-value coal. The processing may require some form of washing, crushing or blending, briquetting or liquefaction. For more upgrading to occur, producers would need much stronger price incentives.
21. The World Bank, "Investment in Flux", *Indonesia Economic Quarterly*, March 2014, p. 21.
22. Jason Allford and Morkti P. Soejachmoen, "Survey of Recent Developments", *Bulletin of Indonesian Economic Studies* 49, no. 3 (2013): 267–88.
23. *U.S. Geological Survey of Metals and Minerals, 2013.*
24. John Kurtz and James Van Zorge, "The Myth of Indonesia's Resource Nationalism", *The Wall Street Journal*, 1 October 2013.
25. Stratfor Global Intelligence, "Indonesia Struggles with an Export Ban", 2014, <http://www.stratfor.com/search/site/Indonesia-struggles-with-an-export-ban> (accessed 26 March 2015).
26. *BP Statistical Review of World Energy*, June 2013.
27. IEA, *Southeast Asia Energy Outlook 2013*, p. 72.
28. Ibid.
29. IEA, *Southeast Asia Energy Outlook*, September 2013, p. 72.
30. Ibid., p. 74.
31. Stratfor Global Intelligence, "Indonesia Struggles with an Export Ban", 2014, <http://www.stratfor.com/search/site/Indonesia-struggles-with-an-export-ban> (accessed 26 March 2015).
32. This paragraph is a summary of "Renegosiasi Berhasil, Harga Jual Gas Tangguh Sesuai Harapan" [Renegotiation was Successful, the Price of Tangguh LNG as We Expected], <http://www.esdm.go.id/berita/migas/40-migas/6862-renegosiasi-berhasil-harga-jual-gas-tangguh-sesuai-harapan.html> (accessed 17 March 2015).
33. *Indonesia Central Statistics Agency Figures.*
34. Michael T. Rock, "What Can Indonesia Learn From China's Industrial Energy Saving Programs?", *Bulletin of Indonesian Economic Studies* 48, no. 1 (2012).
35. In 1982, China's average tariff level for all imports was 56 per cent; in 1992 it was 43 per cent. In 1999 this fell to 15 per cent for all imports and to 8.9 per cent for imports of manufactures. In addition, as part of its accession to the World Trade Organization, China agreed to eliminate all import quotas, import licences and NTBs (non-tariff barriers), and these obligations have by and large been met. See L. Branstetter and N. Lardy, "China's Embrace of Globalization", in *China's Great Economic Transformation*, edited by L. Brandt and T.G. Rawski (Cambridge: Cambridge University Press, 2008), pp. 634–35 and 650–51. While Indonesia's average tariff rate is equally low, NTBs and price controls are widespread, and some local governments have created their own customs agencies. See M. Chatib Basri

and Hal Hill, "Indonesia: Trade Policy Review 2007", *The World Economy* 3, no. 11 (2008): 1,402–4.

36. Ministry of Commerce of China, *2013 Statistical Bulletin of China's Outward FDI.*
37. *ASEAN Statistical Yearbook 2012*, p. 139.
38. Sara Schonhardt, "US, China Vie for Influence Among Indonesian Riches", *Asia Times Online*, 6 May 2011.
39. Readers Forum, "Investment from China's enterprises", *Jakarta Post*, 21 September 2012.
40. According to many scholars, China's policy shift in favour of Asian multilateralism was caused by two factors: the first being the "ASEAN Way" rules of interaction; the second being that with ASEAN in the driver's seat. China felt confident that it would not be turned into an anti-China forum. ASEAN's principles of consensual decision-making, informal diplomacy and non-interference make China feel comfortable sitting around the table with ASEAN. See Fenna Egberink and Frans-Paul van der Putten, "ASEAN, China's Rise and Geopolitical Stability in Asia", *Clingendael Paper No. 2* (The Hague: Netherlands Institute of International Relations, April 2011).
41. Dewi Fortuna Anwar, "An Indonesian Perspective on the U.S. Rebalancing Effort toward Asia", NBR Commentary, 26 February 2013, <http://nbr.org/downloads/pdfs/outreach/Anwar_commentary_02262013.pdf> (accessed 26 March 2015).
42. Makarim Wibisono, "Indonesia and Global Competitiveness", *Jakarta Post*, 10 October 2011.
43. Quoted from Makarim Wibisono, "Indonesia and Global Competitiveness", *Jakarta Post*, 10 October 2011.

6

ENERGY RESOURCE COMPETITION AND THE SOUTH CHINA SEA DISPUTES

China's search for energy resources in Southeast Asia can help deepen China–ASEAN relations in cooperative terms. It can also be a cause for possible conflicts. Rising energy prices, fears of supply scarcity, and rapid increase in oil-import dependence in China and ASEAN countries have helped drive resource nationalism among regional governments and local communities. In the South China Sea, the question of access to maritime energy resources is interlinked with national sovereignty issues, making the South China Sea disputes more complex. This chapter will try to analyse two critical questions: whether the potential energy reserves in the South China Sea has contributed to aggravating these territorial disputes and thus intensified regional suspicions of China's hegemonic ambition; or whether these potential energy reserve can instead provide a means for the parties to seek cooperative and collaborative solutions, and thus improve China–ASEAN relations.

MARITIME ENERGY RESOURCES AND THE "SOVEREIGNTY DILEMMA"

In the South China Sea, maritime energy nationalism is often tied to disputes over territorial claims and fishing rights. Maritime energy resource nationalism is exacerbated by some speculative estimates that the ocean floor is rich of energy resources. Politics, both domestic and diplomatic, supports a predilection for optimistic assessments from the claimant states. Although geographic evidence suggests that the actual prospective areas are quite small, as most of the seabed lacks the characteristics to be seriously prospective,[1] some countries' estimates on South China Sea resource reserves are at a high level. Some international sources have suggested that the South China Sea could be "Asia's Mediterranean", a "new Persian Gulf",[2] and leading Chinese sources give estimates of the potential oil resources that are far higher than other estimates. In November 2012, the China National Offshore Oil Corporation (CNOOC) estimated that the South China Sea held 125 billion barrels of oil (bbo) and 500 trillion cubic feet (tcf) of gas beneath the South China Sea.[3] In January 2013, the U.S. Geographical Survey published a much lower estimate of 11 bbo in proved and probable reserves, and 190 tcf of gas in undiscovered resources.[4] These predictions of rich energy resources have certainly fed the popular perception that oil and gas reserves are at the heart of the territorial disputes in the South China Sea and have strengthened the nationalist sense of the claimant countries' right to defend their claims. The growing importance of energy resources found in the South China Sea has been held responsible for the growing assertiveness of the Chinese government. The ASEAN states depend as much on energy resources as China does, thus escalating territorial disputes in this area.

Southeast Asian claimants are increasingly worried that their traditional approach toward China is being threatened by the growing asymmetry in military power with the expansion of the People's Liberation Army Navy (PLAN).[5] The bilateral relations can easily be undermined by a subsequent rise of nationalist sentiment within claim countries. Sovereignty is, in this sense, a more core national interest than energy supplies and joint development of maritime resources. It is the sense of territorial sovereignty and strategic security rather than energy supply that ultimately drives claimant countries' relations with China with whom they have territorial disputes. This underlines Vietnam's harsh action against the 981 oil rig incident, as it is more concerned about sovereignty than oil.[6] It seems that the Vietnamese government tried to block the project at all costs. The same can be said of China's firm response to the oil and gas extraction by Vietnam and the Philippines which leads to the erection of permanent features that allow de facto control of areas in dispute and later to legal possession, lessening China's claim. Maritime security challenge has become more acute than energy security to Beijing.

At present, China is the largest offshore energy producer in Asia. China's growing offshore energy interests, its growing emphasis on offshore energy production, and its assertiveness in the South China Sea are driving related

ASEAN countries to safeguard their own maritime energy security interests by upgrading their militaries through arms purchases from other big powers. Undoubtedly, any preparations to strengthen a claimant state's strategic position by military means will worry China as this would give them the potential ability to threaten seaborne natural resources and sovereignty issue. These concerns, in turn, could provide China with reasons to hasten its acquisition of long-range power projection systems, such as aircraft carriers and more advance submarines, and to establish air defence identification zones over the South China Sea. In this sense, maritime energy nationalism does have potential to elevate energy disputes and spark armed conflict in the South China Sea.

ENERGY RESOURCE RIVALRY IN THE SOUTH CHINA SEA

The South China Sea has been an important shipping lane for past 2,000 years, though there is little exact textual evidence before the sixteenth century. Its rich fish resources have provided livelihood for the surrounding countries for centuries. But since World War II the discovery of huge oil and gas reserves in the South China Sea "is producing a new geography of conflict in which resource flows rather than political divisions constitute the major fault lines".[7] The positions of the governments of surrounding states have hardened, negotiations have largely failed and a solution is not in sight. The nature of disputes in the South China Sea is much more complex, since strategic interest, national sovereignty, and resource nationalism are key motivating factors along with energy rivalry.

China Adjusting Its South China Sea Policy

From the late 1990s up to only recently, China's approach to Southeast Asia and various territorial and maritime disputes was to engage in a "good neighbour policy" and bilateral negotiations. To alleviate the suspicion and resistance among many countries and create a peaceful international environment for its modernization programmes, China followed a low profile policy and avoided confronting the United States and ASEAN countries while its economy began taking off rapidly in the late 1990s and the early 2000s. For this and other reasons, Beijing tried hard over the time to be benign and charming to Southeast Asia, and adhered to Deng Xiaoping's previous guidance of "shelving the disputes (of sovereignty) and working for joint development".[8]

Recent years, however, found that Beijing's actions had somewhat departed from this general approach and become "assertive" in terms of energy resource exploration and frequent military activities there, adding worries to ASEAN countries. Beijing has renewed its claims and forcefully expanded its maritime law enforcement in the South China Sea by sending "combat-ready" patrol ships regularly to escort fishing fleets and conducting naval exercises in the disputed area of the South China Sea. Moreover, in January 2014, Chinese Hainan province introduced an amended maritime regulation that requires foreign fishing-related vessels to secure the permission of local authorities before entering China's claimed maritime jurisdiction.[9]

For China, energy security and maritime development are the main considerations. The recent crises and turmoil in Sudan, northern Africa, and the Middle East have affected China's overseas energy-strategic areas, posing potential constraints and raising costs. For example, because of the outbreak of the civil war, China's imports of crude oil from Sudan (North and South Sudan) decreased from 13 million tons in 2011 to 2.5 million tons in 2012, a drop of 80 per cent.[10] A large amount of China's investment in energy facilities and infrastructures also got damaged. Moreover, China energy consumption structure mainly relies on coal which has resulted in a series of problems including environmental pollution and climate change. China was propelled to further implement its energy diversification strategy, and shift its oil and gas development focus to the ocean. Moreover, China's maritime capabilities are growing rapidly, including maritime law enforcement, military power projection, and offshore drilling. China has invested considerably in becoming a "maritime power", as the Twelfth Five-Year Plan (2011–15) calls for the national maritime economy to contribute 10 per cent of China's total gross domestic product (GDP).

Domestically, Beijing is facing growing pressure to adjust its South China Sea policy. For example, Yan Xuetong, Dean of the Institute of Modern International Relations, Tsinghua University, pointed out that the South China Sea issue reflects that China's diplomatic philosophy of "non-alliance", "keeping a low profile" is outdated. "Only when China changes its diplomatic principles can it resolve the South China Sea issue."[11] China's accession to the binding Treaty of Amity and Cooperation in Southeast Asia manifests China's self-restraint, while other countries have continued to erode China's maritime rights within the U-shaped line. In this regard, Pang Zhongying, a professor of international relations at Renmin University, noted that "this is a foreign trap designed by ASEAN countries by making use of China's self-restraint policy."[12] Chinese scholars suggested that in the face of big exploration of oil and gas in the South China Sea by other countries, China must accelerate the speed of its development there.[13] Chinese leaders are moderate in their views, but they have to take into account the views of different scholars and the emotional nationalism of young Chinese, especially as expressed on the Internet.

China's Search for Maritime Energy Resources

Under such circumstances, the search for energy resources in the adjacent waters and the support of its oil companies to participate in the development of offshore oil and gas projects have become an inevitable trend for China. Related ministries and departments (oil and gas companies) in China have likewise attached great importance to the development and utilization of oil and gas resources in the South China Sea, suggesting that the South China Sea will undoubtedly become a main source of China's oil and gas supply in the future.

China is currently Asia's largest offshore energy producer, followed by Malaysia, Vietnam, and Indonesia. China's offshore oil production reached more than 600,000 barrels per day (bpd), accounting for about 15 per cent of China's total oil production.[14] Before 2012, China's energy exploration was primarily confined to shallow waters adjacent to its south-eastern coast. Table 6.1 shows the production of CNOOC, China's main offshore oil and gas producer. The Bohai Gulf is presently China's core offshore production zone in terms of oil output. But deep-sea capability has become necessary as China searches for new energy sources to bolster its energy security, a search that has taken place alongside its desire to reshape the political and security environment in the South China Sea. The South China Sea is set to become an important oil and gas source as it is believed that "Beijing sees deep-sea exploration as an important tool to substance China's physical presence in the disputed waters."[15] China is targeting oil and gas production of 500,000 bpd of oil equivalent by 2015 and 1 million bpd of oil equivalent by 2020 in 3,000 metres deep-sea areas of the South China Sea.[16]

To accompany Beijing's strategy, the last five years has seen CNOOC shift its focus toward developing its deep-sea exploration capability. Though still in its early stages, CNOOC has partnered with foreign companies

TABLE 6.1

CNOOC's Oil and Gas Production in Adjacent Waters

	2005	2006	2007	2008	2009	2010	2011
Net production of crude and liquids (barrels/day)							
Bohai Bay	178,840	200,944	206,748	218,478	253,884	408,946	405,682
Western South China Sea	49,016	40,437	34,163	56,761	72,605	84,116	72,006
Eastern South China Sea	103,741	105,902	103,715	122,813	118,391	121,454	120,563
East China Sea	1,706	1,464	1,467	85	63	53	339
Overseas	23,565	23,973	25,735	23,931	64,749	90,419	83,993
Total	356,868	372,720	370,433	422,068	509,606	704,988	682,583
Net production of natural gas (million cubic feet/day)							
Bohai Bay	49.1	64.5	70.2	74.5	79.2	120.4	123
Western South China Sea	229.6	251.8	237.3	284.7	275.4	354	390.4
Eastern South China Sea	—	23.1	27.4	28.1	50.2	139.5	157.8
East China Sea	18.3	21.2	8.7	6.8	6	5.5	18.7
Overseas	92.7	130.3	200.7	227	242.7	332.2	345.3
Total	389.6	490.9	544.3	621.1	653.5	951.6	1,035.2

Source: *CNOOC Annual Report 2010*, 2011, <http://www.cnoocltd.com> (accessed 11 January 2014).

and increased indigenous development to expand its technological reach. In May 2012, CNOOC began its first deep-sea project in an undisputed area of the South China Sea, south-east of Hong Kong. Down Erica, an American scholar with the Brookings Institute, believed the deployment of CNOOC's new rig indicates that CNOOC is beginning to close the gap with major international oil companies in the area of deep-water drilling.[17] Currently, CNOOC is equipped with two deep-sea oil platforms, CNOOC 981 and Nanhai VIII, which will drill in water depths of up to 3,000 metres (9,800 feet) and 1,400 metres, respectively.[18]

ASEAN Countries' Exploration Activities

China is not the only country which is assertive in exploiting energy resources in the South China Sea. In fact, Vietnam, the Philippines, Malaysia, and Singapore were already exploiting oil resources after they were discovered in the South China Sea in the 1960s. There were hundreds of platforms in the Spratly Islands area, and according to China's data, more than 50 million tons of oil is extracted per year.[19] Vietnam and the Philippines are the most active in their exploration activities by seeking cooperation with international oil and gas companies.

Vietnam

Vietnam is a major offshore oil producer in the South China Sea. In 2011, its state-owned oil company Petro Vietnam, produced 30 million tons, or 27 per cent of Vietnam's total production from three fields in the South China Sea.[20] What Chinese experts on Vietnam are more concerned about is Hanoi's efforts to develop a comprehensive ocean strategy.[21] The 4th Plenum of the Tenth Vietnam Communist Party's Central ommittee (VCPCC) held in January 2007 endorsed an "Ocean Strategic Program to 2020", reflecting ambitions to make Vietnam a "maritime major power". By 2020, the goal was for Vietnam's oceanic economy to provide 53–55 per cent of its GDP, with ocean exports amounting to 55–60 per cent of total exports.[22] To realize these ambitions, Vietnam has had an incentive to involve as many foreign partners as possible to reinforce its claims in the area and to deter Chinese opposition.[23] The most striking evidence is the fact that in 2013, U.S.-based ExxonMobil and PetroVietnam announced plans to build a US$20 billion power plant to be fuelled by the oil and gas from two exploratory areas, which are west to Paracel Islands but within Vietnam's exclusive economic zone (EEZ).

Another country which has increasingly got involved in oil and gas exploration with Vietnam is India. ONGC Videsh (or OVL, for short) is the subsidiary of India's Oil and Natural Gas Corporation (ONGC), a state-owned company under the Ministry of Oil and Natural Gas. It had been present in Vietnam for quite some time, including in major oil ventures for offshore oil and gas exploration. India already has a stake in a block located 370 km south-east of Vung Tau on the southern Vietnamese coast with an area of 955 square km. The exploration licence for this block was acquired by OVL in 1988. The field started commercial production in January 2003. During 2010–11, OVL's share of production from the project was 2.249 billion cubic meters (bcm) of gas and 0.038 million metric tons of condensate.[24]

Later in 2006, OVL acquired two more blocks (Block 127 and Block 128) in Vietnam's EEZ in territories under dispute for hydrocarbon exploration. OVL had invested around US$68 million by March 2010, but withdrew from Block 127 which proved unviable and dry.[25] While Block 128 was bogged down by layers of hard rock and unfavourable geological conditions which made it difficult to penetrate. Despite these issues, India decided not to withdraw from Block 128 due to complex geo-strategic reasons, including a request from the Vietnamese to stay for another two more years.[26]

In spite of China's protests against the exploration activities of OVL around the Paracel Islands, OVL took the view that Vietnamese claims are in accordance with international law, and it will continue with exploration projects in the two blocks near the Paracel Islands.[27] In October 2011, a three-year agreement for cooperation in oil and gas exploration and production was concluded between ONGC and PetroVietnam. There was also an Memorandum of Understanding signed, in which the seven oil blocks in the South China Sea were offered to India (including three on an exclusive basis) as well as joint prospecting in some Central Asian countries with which both Hanoi and New Delhi have good political ties.[28] According to Ghosh, an Indian observer, "Vietnam's granting of these seven oil blocks in the South China Sea to India for exploration is part of a plan to internationalize Hanoi's territorial dispute with China."[29]

The Philippines

For the Philippines, the urgency of developing marine resources is great as well. From the perspective of per capita resources, the pressure it faces is more serious than China. The Philippines' population density is 342 persons per square km, while China is 140. Therefore, in the Philippines there is general consensus on the importance of marine resources for its national economy.

In terms of energy supply and demand, according to the International Energy Agency (IEA), the Philippines' primary energy demand stood at 40 Mtoe (million tons of oil equivalent), and is projected to grow at an average rate of 3.5 per cent per year from 2011 to 2035.[30] Oil accounted for 35 per cent of the Philippines' total primary energy consumption in 2011. The country's oil consumption in 2012 was 282,000 bpd, but its production in 2008 was only 23,000 bpd, most of which was from the Malampay and Palawan fields in the South China Sea.[31] Moreover, due to domestic political and religious reasons, the Philippines' relations with Arab countries and Indonesia are complex, leading to its unstable energy supply. Therefore, it wants to expand its domestic oil production to reduce its almost total reliance on oil imports.

The Philippines has ambitious plans for its energy strategy and has attempted to boost self-sufficiency in oil production. It had intended to offer fifteen exploration contracts for offshore exploration off Palawan Island in an area claimed by China.[32] One of the prominent cases is that the Philippines is seeking to develop the seabed hydrocarbon resources of Reed Bank in the South China Sea, an area under dispute with China. On May 2014, the Philippines' Department of Energy launched a tender for exploration rights to eleven oil and gas blocks, and Philex Petroleum Corp of the Philippines announced that its London-listed unit, Forum Energy Plc, plans in early 2016 to start drilling appraisal wells in the Sampaguita gas field, also at Reed Bank.[33] Should the Philippines proceed to develop Reed Bank unilaterally, there is a likelihood that tensions between the two countries will escalate.

In China's view, these "joint developments" among different countries has compressed its geo-economic space in the South China Sea, as Chi Fulin, Dean of China (Hainan) Reform and Development Research Institute, says:

> a wide range of oil and gas wells of other countries have indirectly claimed sea area of more than 1.5 million square kilometers, all of which is disputed by China and peripheral countries around the South China Sea; consequently, only sea area of 440,000 square kilometers is left for China, which only accounts for 22 per cent of China's claimed sea area within the South China Sea.[34]

Moreover, Chinese scholars believe that this long-term joint development projects will easily cause a "fait accompli" for China's geo-economic strategic space in the South China Sea, and thus increase the vulnerability of China's geo-economy in that area.[35]

External Players in the South China Sea

As tensions in the South China Sea increased, some external players such as the United States, India, and Japan are also increasingly getting involved in the South China Sea dispute, either for their own energy and strategic reasons, or for ASEAN countries' needs or as a hedging strategy to balance China's influence.

The United States

For a long time, the U.S. position on South China Sea disputes was neutral, insisting that it did not take sides in territorial disputes and had no stake in them. The United States supports the ASEAN initiatives with respect to the Declaration on the Conduct of Parties in the South China Sea (DOC) and Code of Conduct (COC). In 2002, ASEAN and China agreed on the DOC which was a set of principles that was supposed to stabilize the status quo, though it was non-binding and lacked any enforcement mechanism. ASEAN's 2011 leader, Indonesian President Susilo Bambang Yudhoyono, stated at the Association's 44th Ministerial Meeting in July that the ASEAN Regional Forum (ARF) should "finalize the long overdue guideline because we need to get moving to the next phase, which is identifying elements of the COC". The United States insists all parties accelerate efforts to agree on a full COC for the South China Sea.

Beginning with the ARF in July 2010, the Obama administration decided to play a major role in promoting resolution of the Spratly Islands imbroglio, while laying down a marker that the South China Sea stability for maritime commerce constituted a significant U.S. interest.[36] At the July 2010 ARF, U.S. Secretary of State Hillary Clinton affirmed this:

> The U.S., like every other nation, has a national interest in freedom of navigation, open access to Asia's maritime commons, and respect for international law in the South China Sea. We share these interests not only with ASEAN members and ARF participants but also with other maritime nations and the broader international community.[37]

This suggested that on the South China Sea issue, the United States has adopted a similar stance in relation to the Diaoyu/Senkaku dispute between China and Japan in the East China Sea, allowing also for the security commitment it has in relation to Japan.

The United States holds that territorial disputes must be reconciled according to international law. This means the 1982 United Nations Convention on the Law of the Sea (UNCLOS), which has rules for fixing maritime boundaries via EEZs. Application of these principles would invalidate China's claims to most of the South China Sea as it favours the littoral states. In addition, Washington has become a strong backer of ASEAN's multilateral negotiation posture, strongly urging China to compromise in multilateral talks with its ASEAN counterparts to find a solution in the South China Sea disputes. But Beijing's strong preference is to work through bilateral talks. After all, the United States has a security treaty with the Philippines which could potentially draw it into the conflict. The United States' emphasis on multilateral diplomacy for the South China Sea underlines the hope that ASEAN, as a whole, gets more involved and plays a bigger role in efforts to resolve the disputes. Actually, within ASEAN, the features claimed by Malaysia, the Philippines, and Brunei are also claimed by Vietnam. So not only are these claimants arrayed against China but also against each other. Moreover, ASEAN states take varying positions on the South China Sea disputes: Laos, Cambodia, and Myanmar lean toward China; Malaysia and Indonesia are cautious about U.S. involvement; Thailand and Singapore are neutral; while both Vietnam and the Philippines welcome an American role.[38] Hence, it is not surprising to see that the diplomatic tit-for-tat over this may cause a division of the ASEAN countries.

Japan

Japan is another big power getting involved in the South China Sea issue. Though a non-claimant state in the South China Sea, Japan has great concern about the dispute. The Japanese perceive a link between the South China Sea and East Asia Sea disputes, and that Beijing's strategy and actions toward the claimant states in the South China Sea may have implications for the East China Sea and the Diaoyu/Senkaku dispute. Thus, in seeking a more active political role in the South China Sea disputes, Japan is strengthening its diplomatic and defence ties with the Philippines and Vietnam, and using multilateral institutions like the ARF and the East Asian Summit (EAS) to check perceived Chinese assertiveness in the South China Sea. In September 2011, Japanese and Philippine officials discussed the creation of a "permanent working group" to coordinate their policies pertaining to Asian maritime disputes.[39]

As the South China Sea disputes got tense in 2014, with China's Haiyang Shiyou 981 oil rig being deployed in disputed waters, Prime Minister Shinzo Abe pledged at the Shangri-La Dialogue held in Singapore on 30 May 2014 to support Vietnam and the Philippines in their territorial disputes with China, saying that Japan will provide patrol vessels to these two countries.[40] This shows that PM Abe's Japan is ready to assist Southeast Asian states with claims to the South China Sea to monitor and fend off China, and is determined to become a regional security leader or co-leader for Southeast Asia.[41] The easing of the arms export restrictions in 2011 may have played a role in the pledge to transfer these vessels for the purpose of helping the Philippines and Vietnam in enhancing their maritime security. Obviously, the Philippines and Vietnam are in such a bind in their territorial disputes with China that they will grasp any military assistance from Japan. But it is harder for some other ASEAN states like Malaysia and Indonesia, as they are likely to be acutely aware that to back PM Abe's vision for Japan will really upset their relations with China.[42]

Besides strengthening ties with the Philippines and Vietnam, Japan has always sought to use the ARF and EAS to deal with the South China Sea dispute. For example, in October 2011, Japanese Foreign Minister Gemba Koichiro raised a proposal for a maritime regime to be floated at the EAS meeting to be held in Bali, Indonesia the following month.[43] The key Japanese idea was a multilateral approach which called for Japan and ASEAN's dialogue partners to construct a maritime regime in the South China Sea based on the freedom of navigation, international law, and peaceful settlement of disputes. Not surprisingly, Tokyo's attempts to build a new maritime architecture for the South China Sea has been viewed as "muddying the water" by Beijing.[44] The more Japan confidently inserts itself into the South China Sea disputes, the more Tokyo's enmity with Beijing becomes a part of the wider picture.

India

India is not a party to the South China Sea dispute, but it is getting more and more interested in the region. India's strategic expansion from the Indian Ocean into the South China Sea may be understood along several dimensions. First, it desires to become an Asian power, not just an Indian Ocean actor. Second, India, after considerable investment in its navy, now has the capability to deploy to eastern Asia and balance China — not only from along the Sino–Indian land border but on the sea as well. Third, India is investing in the South China Sea energy exploration for its rapidly developing economy. With China–ASEAN ties under stress due to Beijing's territorial claims, New Delhi has been trying to fill the void by emphasizing its credentials as a responsible regional stakeholder. The most striking evidence that India is interested in South China Sea energy resources is its joint-exploration with Vietnam.

NEGATIVE IMPACTS OF THE SOUTH CHINA SEA DISPUTES

Southeast Asian Perceptions

From Southeast Asia's point of view, there have been mixed signals coming from Beijing in recent years. On the one hand, China accepted guidelines on how to implement the 2002 DOC suggested by ASEAN, according to which all parties pledged to seek peaceful solutions to disputes and conduct maritime cooperation in order to maintain regional stability in the region. On the other, recent years have also seen China grows assertive in terms of energy resource exploration, island expansion, and military activities in the South China Sea. ASEAN recognizes that China prefers to use its size to dominate its regional relationship by focusing on bilateral ties or on regional structures that China can more readily influence. This may destroy the mutual trust and add worries to ASEAN claimant countries, resulting in ASEAN becoming more pragmatic by emphasizing a hedging strategy with regard to China.

The Philippines

Among ASEAN countries, the Philippines was particularly affected by threat perceptions arising from China. In many Filipinos' view, China's political system is seen widely divergent from their own. Decades of

Cold War anti-communist ideological posture have contributed to a high degree of mistrust of the Chinese Party-State. The perception of China as an atheist society by Filipinos, many of whom find religiosity and spirituality integral to their identity, also adds to the ideological distance and anxiety about China.[45] The Philippines perceives China's assertive stance on territorial and maritime jurisdictional disputes in the South China Sea, demonstrated especially in military terms, as a threat. The Philippines believes that China's control of Mischief Reef and other reefs after the late-1990s constitutes a threat to its national security, and is thus impelled to rely on the U.S. military to balance China's behaviour in the South China Sea.[46] In October 1995, former Philippines President Fidel Ramos delivered a speech at the East-West Center in Hawaii postulating that China posed a threat to regional security and calling on the United States to retain its military presence in Asia. He said "even if Beijing does not have the capability to expand beyond its borders, China will inevitably be a political and military threat to Southeast Asia".[47]

Today, due to territorial disputes escalation in recent years, there is still outright anxiety and concern about China's strategic role in the region, a degree of wariness toward this major power is more evident among ordinary people. For example, the Pew Research Global Attitudes Survey released in 2013 showed that China's favourability in the eyes of the Philippine respondents decreased from 63 per cent in 2002 to 48 per cent in 2013, and only 22 per cent of Filipinos think of China as a partner, while America scored better at 81 per cent.[48] Renato Cruz de Castro, an expert on the Philippines, believes that "as a close neighbor of China, the Philippines has not yet totally trusted Beijing. Manila still considers Washington as the least dangerous among the big powers, the best balancer, and the most reliable insurance against an emerging China."[49] Philippine Foreign Minister Albert Del Rosario also stressed that "U.S. is the sole strategic partner of the Philippines".[50] From security and strategic perspectives, being the smaller and weaker parties to the South China Sea disputes, the Philippines and Vietnam have strong incentives to solicit for American support to counter China's expansion there. "The Philippines has no choice but to rely on ASEAN and redefine U.S.-Philippines relations."[51]

Subsequently, in line with the Obama administration's "return to Asia" strategy, the Philippines has been trying to demonstrate that it accords a high priority to relations with the United States. At the height of Manila's territorial disputes with China in the first half of 2012, the United States has sent a strong signal of support for the Philippines, speaking out on the issue at several ASEAN forums, selling the Philippines a decommissioned Hamilton-class coast guard cutter in October 2011 and promising another one, increasing troop rotations and joint training in the Philippines, and committing to expanding port visits and joint exercises between the navies of the United States and the Philippines. All these have led to a new plateau of distrust and tension in Philippine–China relations.

Vietnam

Traditionally, in response to China's behaviour in the South China Sea, Hanoi had adopted "a two-pronged policy of 'hedging': that is, pursuing engagement along with indirect balancing and trying to maintain balanced relations between powers without firmly plumping for either."[52] In other words, different from some other ASEAN countries, Vietnam was using its traditional approach to deal with China — stressing commonalities and friendship while seeking counterbalance from other powers. However, China's recent hardline approach on its territorial disputes has pushed Vietnam toward a more active hedging strategy, and this hedging has been perceived as an opportunity by the Obama administration for its "return to Asia" strategy.

Although Vietnam-U.S. relations cannot go much further beyond the constraints imposed by both Vietnamese concern of Chinese reactions and the U.S. Congress which has hindered the government's effort to build closer ties with Vietnam, a number of notable visits have taken place in recent years. In August 2011, the two countries concluded their first military agreement since the Vietnam War; though this was limited to cooperation in health and research collaboration in military medicine, it is likely to open the door to other and wider agreements.[53] In June 2012 when visiting Vietnam, U.S. Defense Secretary Leon Panetta urged Vietnam to host more U.S. military craft "as the U.S. would shift emphasis to Asia by working with partners like Vietnam".[54]

In early May 2014, China deployed the drilling platform Haiyang Shiyou 981 in disputed waters of the South China Sea, causing a wave of protest in Vietnam. After the oil rig deployment incident, in addition to soliciting support from ASEAN, Vietnam has been trying to enhance ties with important partners such as the United States, the Philippines, and Japan. The most symbolic act was its decision to participate in the Proliferation Security Initiative (PSI) after more than ten years of consideration since the initiative was launched in 2003. Vietnam had been reluctant to support the initiative, which it sees as outside the United

Nations' framework. The announcement by Vietnam's foreign minister that his country will join the PSI, combined with its commitment to enhance cooperation with the United States in maritime security, and the U.S. promise of providing US$18 million of aid to the Vietnamese coast guard, signifies an increased level of mutual trust in the Vietnam-U.S. security cooperation.[55]

Indonesia

Jakarta was unaffected by the territorial disputes in the South China Sea in the 1990s. Nevertheless, Indonesia, which perceives itself as the leader of the region, was concerned about the potential of the South China Sea issue to affect regional political stability. In March 1995, conflict over Mischief Reef in the South China Sea between Beijing and Manila led to renewed concern over Beijing's intentions. Indonesia was concerned that China's claim in that area might also infringe upon Indonesian sovereignty over Natuna Island. In this context, Indonesia had expressed its concerns over the rise of China's military capability and how China would use it in the future.[56] Jakarta soon began to pay attention to the Natuna islands. This was especially so after China was reported to have included the oil-rich area into a map detailing its claims in the South China Sea.[57]

In more recent years, as Beijing's attitude toward the South China Sea issue has been hardening, there was a new dimension to the Indonesian view: worry about China's regional role and policies, especially regarding Beijing's behaviour in the South China Sea disputes. Indonesia was the only ASEAN country that told Beijing earlier this year that Jakarta would not accept a Chinese air defence identification zone over the South China Sea. In a significant policy shift, Indonesian officials on 12 March 2014 announced that China's nine-dash line map outlining its claims in the South China Sea overlaps with Indonesia's Riau Province, which includes the Natuna Island chain.[58] "Indonesia's public declaration that it has a maritime conflict with China is a potential changer in the game being played out in the South China Sea."[59]

At stake for Indonesia is not only the Natuna islands and surrounding waters, but also the sanctity of UNCLOS. Indonesia is the world's largest archipelagic state and it lacks the naval capacity to defend its far-lung archipelago, which spans 4,800 km from east to west. It has therefore always been a strong advocate of UNCLOS. In Indonesia's view, in recent years, China has taken a series of actions that Indonesia perceives as undermining UNCLOS and threatening regional stability. First there was China's 2009 publication of its nine-dash line map, which includes parts of the Natuna island EEZ in its southern-most area. Indonesia protested China's claims to UNCLOS in 2010, and also requested that China clarify its claims by providing precise coordinates. Second, China has recently become assertive in pursuing its claims. Most critically from the Indonesia perspective, China has expanded its naval excercises from its northern claims closer to mainland China down to its southern ones. In 2010, for example, after an Indonesian patrol boat captured a Chinese vessel fishing within its EEZ, the Chinese dispatched the *Yuzheng 311*, compelling the Indonesian patrol boat to release the Chinese vessel. Similarly in March 2013, Indonesian officials boarded a Chinese vessel fishing in the Natuna Islands and transferred the Chinese crew to its boat to be taken ashore. Before reaching land, Chinese armed vessels confronted the Indonesian boat, and demanded the release of the Chinese fisherman.[60]

Thus, although Indonesia has reservations about external powers' role in regional security, it has changed this attitude, supporting the United States' deployment of U.S. marines in northern Australia. Indonesian Foreign Minister Marty Natelagawa rejected China's view that the United States should not become involved in the South China Sea dispute.[61] Nevertheless, Indonesia's support for the United States's involvement is limited, and it is unwilling to see external powers deploy too many forces in the region, as Indonesia's ambition is to become the leader of the region.[62]

Malaysia and Brunei

Malaysia and Brunei, the other two claimant states, have studiously adopted a low public profile on the South China Sea. Malaysia believes that "China is a country you can expect and believe",[63] and is willing to solve the problem through bilateral diplomacy, supporting China's proposition that the South China Sea issue should not be internationalized.[64] China also holds that "China and Malaysia are sincere friends who trust and support each other, and are reliable partners who cooperate equally and mutual beneficially."[65] This was reflected by the realities that Chinese fishing boats can regularly go into Malaysia's EEZ without confrontation, and Chinese paramilitary vessels regularly watch vessels operated by Petronas, the state oil company, servicing off-shore rigs in Malaysia's EEZ.

In 2013 and January 2014, a PLAN flotilla travelled to James Shoal, 80 km off the coast of East Malaysia and the southernmost point of China's nine-dash line claim to the South China Sea. The chief of Malaysian Armed Forces confirmed the Chinese flotilla had been monitored as it "strayed into Malaysian waters. As long as it is an innocent passage, that is okay with us." Malaysian officials privately stated that the "see nothing, know nothing" stance is dictated by Prime Minister Najib Razak who controls South China Sea policy.[66] Malaysian officials are aware of Chinese fishing activities and other assertions of Chinese sovereignty in the EEZ. In 2013, for example, Malaysian diplomats privately briefed academics from an ASEAN think-tank and told them that aerial photos confirmed that PLAN flotilla near James Shoal.[67]

Nevertheless, as China is getting assertive recently, Malaysia is also concerned that the approach of bilateral negotiations is not conducive to the side of Malaysia and ASEAN. Therefore, although Malaysia tends to resolve the border dispute through bilateral consultation and cooperation, it prominently emphasizes the integrity of ASEAN, promoting the discussion of the South China Sea issue through multilateral talks and forums and developing multilateral relations with countries outside the region. In June 2011, at the 10th Shangri-La Dialogue meeting held in Singapore, Malaysian PM Najib pointed out that

> China is our [ASEAN] partner, the U.S. is as well. I want to explain to our friends from the U.S., China, Russia, India as well as other regions that in ASEAN, we have common values and aspirations. We invite you to have positive cooperation with us. For the future, we need multilateral relations rather than bilateral relations to replace the bilateral relations during the Cold War.[68]

The Malaysian government also responded positively to the United States's South China Sea policy, expressing its understanding of the U.S. military forces stationed in that area.[69]

From the Chinese Perspective

Although ASEAN countries remain uncertain about China's long-term intentions in Southeast Asia, they realize that the economic importance of China has grown, while worrying that U.S. trade and economic policy is ideological and inconsistent with its geostrategic objectives.[70] As a consequence of China's economic rise and diplomatic assertiveness, most ASEAN countries have China as their major trading partner and foreign investor while they depend on the United States for the maintenance of the regional security order. This strategic "dual dependency" on the United States and China has led to the strategic ambivalence in ASEAN as a regional organization and as individual countries. This has affected their economic cooperation with China to a significant extent.

Take the Philippines as an example. Throughout most of the 1980s, the Philippines was beset by political instability and economic malaise, and was in dispute with China on the South China Sea issue. Therefore it was not in a position to take advantage of China's economic liberalization. The country fell behind most of its ASEAN neighbours, whose trade and investment ties with China expanded remarkably during the period. The situation did not change until 2000 when the former Philippine President Joseph Estrada visited China, signing a joint declaration on the Framework of Bilateral Cooperation in the 21st Century. Since then, China–Philippines bilateral trade rose rapidly. From 2003–11, bilateral trade rose steadily from US$9.4 billion to US$32.3 billion, an increase of 244 per cent, making China the Philippines' third largest trading partner after the United States and Japan.[71]

However, most Filipinos tend to judge Chinese foreign policy on the single issue of how it handles its dispute with the Philippines in the South China Sea, although China's foreign policy goals and interests have a much wider reach.[72] Thus, although economic ties have continued to improve, the degree of wariness toward this major power increased as the tension in the South China Sea heightened, and the fears have extended to economic areas. In the Philippines, there was a great fear that the asymmetrical economic interdependence between these two countries could spell trouble for the Philippine economy. The Philippines still has distrust and uneasiness about Chinese investment, and has yet to jump completely on to the Chinese economic bandwagon. In the Philippine perspective, "Beijing's main motive for developing economic relations with the Philippines is to wean it away from Washington, while isolating the U.S. politically and diplomatically to the maximum possible extent."[73]

Thus, in contrast to other ASEAN countries, the Philippines has not been very active in negotiating Free Trade Agreements (FTA) with China and other countries. Manila did not have clear strategies or policies towards ASEAN–China FTAs and has largely been a follower of the trade negotiations in ASEAN.[74] Consequently, compared with other Southeast Asian countries, the growth in the Philippines' trade with China has been much slower in the past years.

For example, in 2013, the Philippines' trade with China was lowest among the ASEAN-5, only about one third that of Malaysia's trade with China. China's overall investment in the Philippines also remains comparatively small. In 2012, China's foreign direct investment (FDI) in the Philippines was only US$74 million, accounting for less than 1 per cent of China's total FDI in ASEAN in 2012.[75] The tactical decision made in late 2013 to revise Manila's strategy toward Beijing has been evident in President Benigno S. Aquino's repeated calls for direct talks with the Chinese leadership. Indeed, how to deal with the opportunities and challenges posed by a rising China is the key policy issue to be faced by the Philippine government. After all, even the United States itself has realized that it must make use of China's capital and funds to help maintain its huge deficit and strategic ambition. As stated in Ian Bremmer's cover story for *The National Interest:* "In 1977, China accounted for just 0.6 percent of global trade, and in 2012, it became the world's largest trading nation. Today, 124 countries count China as their largest trade partner, compared to just 76 for the U.S.".[76] The Philippines may, in the near future, find it difficult to adjust its relationship with the rising China.

The impact of the South China Sea dispute on China–Vietnam economic ties may be reflected differently. Compared with the Philippines, the dynamics in the economic relations between Vietnam and China are rather different. Hanoi's strategies are shaped by its history, economy, and geographical proximity with China. Hence it holds a more positive attitude to economic cooperation with China, more willing to separate territorial disputes with economic issue. In Vietnam, Chinese FDI increased from US$189.2 million in 2011 to US$350 million in 2012, with the cumulative amount being US$1.6 billion. In fact, as a large proportion of Vietnam's FDI inflows originate from Hong Kong and the British Virgin Islands, a considerable proportion of Chinese FDI in Vietnam was realized through Hong Kong.

In Vietnam, China started with the labour-intensive manufacturing industries, which largely targeted the global export markets. Chinese investment in Vietnam covers a variety of fields, spreading over a range of industries from garment and electronics to motorcycle manufacture. Obviously, Vietnam has a comparative advantage in labour forces, mineral and agricultural sources; while China has a comparative advantage in industrial products, capital and technology. These economic complementarities show that there is substantial intra-industry trade made by multinational companies. In this intra-industry trade, China exports capital goods and industrial intermediate products to Vietnam to exploit cheap labour and raw materials there, while Vietnam exports finished products to the United States and the European Union. The United States is actually the largest export market of Vietnam. From 2005 to 2011, Vietnam's exports to the United States also increased by nearly two times, from US$5.9 billion to US$17 billion as the International Monetary Fund (IMF) figures show. The European Union is the second largest export market of Vietnam. From 2005 to 2011, Vietnam's exports to the European Union increased more than two times, from US$4.1 billion to US$12.6 billion.[77] This fact would suggest that the labour division based on comparative advantages between China and Vietnam has made both sides closely linked in the international production network, forming an important part of the global supply chain. In this sense, any disruption of the China–Vietnam bilateral economic ties caused by the South China Sea disputes may not only affect Vietnam's industrialization process, but also pose a threat on Asia-Pacific trade and the global production chain.

"JOINT DEVELOPMENT" AS A WAY FOR REDUCING TENSIONS AND ENHANCING STABILITY?

While it is accepted that maritime resource competition and territorial disputes in the South China Sea can lead to conflict and harm China–ASEAN relations, the role that maritime resource cooperation plays in ensuring regional stability and peace-making should not be ignored. The fact that China has emerged as an increasingly large regional investor and energy resource consumer, and the emphasis it puts on getting as much of its future oil and gas from as close to home as possible, is increasing its interest and incentives in developing energy resources in the South China Sea. ASEAN countries, on the other hand, are also turning to cleaner burning gas to generate electricity and create a clean environment. Thus, energy resource rivalry might create potential conflicts as well as cooperation opportunities. While maritime territorial disputes are known to be a major irritant in interstate relations, it is regularly argued that joint development provides a means to remove this irritant, albeit temporarily, in a way that does not compromise the claims or positions of the parties.[78]

The prospect for joint development of energy resources in the South China Sea has been under discussion since the early 1990s. It has been China's proposal to pursue joint development and to "shelf" the territorial dispute in 1978 over the Diaoyu/Senkaku islands. Similarly, in the South China Sea, China has incorporated the offer of joint energy development as an integral part of its strategy of reassurance toward Southeast Asia since the early 1990s. Chinese Premier Li Peng stated in 1990 that Beijing was ready to shelve the issue of sovereignty in favour of joint development in the South China Sea.[79] In 2004, China agreed to joint seismic investigations of underwater resources with the Philippines in the South China Sea, and Vietnam joined this research project in 2005. These developments followed on from a series of Chinese commitments that have sought to reassure ASEAN countries of its intentions, which include the 2002 ASEAN declaration on the Conduct of Parties in the South China Sea aimed at preventing the escalation of ongoing maritime disputes over the Spratley islands and the 2003 signing of the Treaty of Amity and Cooperation between China and ASEAN.

However, the proposal to "pursue joint development and shelve disputes" does have its limits. The possibility of joint development relies on a critical precondition that there is consensus on areas in dispute that may be subject to joint development arrangement. Yet, the following two cases show that for the claimant countries, the sovereignty and security considerations may well be above the joint development of energy resources. Hence, as Beckman, a law expert, notes, joint development agreements may not be concluded in the South China Sea as long as the claimants have not clarified their ambiguous claims in conformity with UNCLOS.[80]

In 2011, Philippine President Benigno S. Aquino raised a proposal for joint development called the Zone of Peace, Freedom, Friendship and Cooperation (ZoPFFC). The proposal advocated "a multilateral, ruled-based approach to the disputes, in contrast to China's preference for bilateral talks".[81] It aimed to "separate" disputed from non-disputed areas in the South China Sea as a starting point in negotiating joint cooperation and development zones. ZoPFFC called for the establishment of "enclaves" of the disputed reefs and the establishment of a joint development agency. Furthermore, under ZoPFFC, the joint exploration of resources would be excluded within non-disputed areas, as states would instead be allowed to develop resources unilaterally within their defined EEZs and continental shelves. Beijing quickly rejected the ZoPFFC proposal because the maritime areas claimed by the Philippines to be non-disputed fall within the U-shaped line.[82]

In February 2012, President Aquino also rejected a Chinese offer to jointly explore energy resources in the Areas (Reed Bank off Palawan Island) that the Philippines does not consider to be disputed with Beijing, and Manila insists that Chinese participation in related projects should be in accordance with the Philippines Constitution, with recognition of the area as belonging inside the Philippines' EEZ, and subject to the laws of the Philippines.[83]

For the failure of joint development, there are many different analysis and explanations. From the Southeast Asian perspectives, Beijing has not suggested that the "shelving" of the territorial disputes and the promotion of joint development means that their sovereign claims have become less strong or that such joint development would lead to a longer-term prospect of territorial compromises,[84] as China had indicated that "Beijing would only concede to joint cooperative activities if the other claimants first acknowledge Chinese sovereignty over the South China Sea."[85] Critics charged that joint development was a convenient means for Beijing to claim territory over which it is widely perceived to have little plausible legal claim.[86] Hence, the claimant states have continued to argue over the sovereignty issue instead of temporarily shelving these questions to the benefit of the establishment of a joint development scheme.

From the Chinese perspective, however, domestic political changes and other big powers' involvement are more important factors in affecting the implementation of joint development projects. The United States pivot to Asia takes the form of picking sides in the South China Sea disputes and encourages some claimants to stand up to China; a regime change in Manila in 2010 visibly altered its Spratly approach vis-à-vis Beijing.[87]

In 2004, Manila signed with Beijing the "Joint Maritime Seismic Undertaking Agreement" when the Philippines President Gloria Arroyo visited China in 2004. Different from the ZoPFFC proposal, the Joint Maritime Seismic Undertaking Agreement had allowed for joint seismic surveys to be conducted within the Philippine 200 nautical mile EEZ,[88] and was once considered to be specific measures to "change the South China Sea from a sea of dispute to a sea of peace, stability, and cooperation" and "a break-through" in Philippines–China relations.[89] Unfortunately, during the later period of her presidency, President Arroyo got involved heavily in corruption scandals and was considered as the Philippines's "most unpopular president" since the restoration of constitutional democracy. Sino–Philippine relations were therefor badly affected and those South China Sea cooperation projects were "stigmatized" as a result.[90] Under such a domestic political background, the Joint Maritime Seismic Undertaking Agreement could not get renewed after its expiration in 2008. In February 2009, the Philippine Senate enacted a law that included the Spratlys within the country's maritime baselines.

According to Chinese scholars, the reason for Sino–Philippine joint development projects in the South China Sea to have met strong opposition from domestic society is largely due to the "corruption image" of the Arroyo government among its people.[91] Indeed, before the Arroyo government was linked with the problem of corruption, domestic nationalist sentiments did not give much influence to the Philippines' South China Sea policies. Since the restoration of constitutional democracy in 1986, the Philippines has experienced long-term internal instability and faced the internal security threat of various insurgencies. Against such a background, the Filipinos were more concerned about domestic political issues rather than external issues.[92] And in those domestic issues, the corruption problem was the one which most easily made people be dissatisfied with the government. Corruption problems were also considered as one of the main reasons for hampering the Philippine economic development.

The report of the International Crisis Group noted that compared with Vietnam, the Philippine government was less affected by the nationalist sentiment in dealing with the South China Sea issue. In contrast, on the relationship with the United States, the Philippine nationalist may create troubles to the foreign policymakers.[93] Rand Corporation also noted in its released report in 2008 that Sino–Philippine relations seldom appear as a major topic in the Philippines politics. The two topics that have attracted the most attention are the "U.S.–Philippine defence relations" and the "Filipino overseas workers". Actually, the making of China policy is more influenced by the political elite rather than mass politics. The specific policies toward China are mainly decided and made by the president himself and the core think-tanks.[94]

During the Philippine 2010 election, Benigno S. Aquino III played the campaign slogan of "no corruption, no poverty", illustrating his understanding of public opinion. After taking office, President Aquino was determined to establish a clean image different from President Arroyo's so as to win popular support. Meanwhile, he launched the "liquidation" campaign against President Arroyo and her allies, hence Arroyo's South China Sea policy naturally became President Aquino's attacking target.[95] This is considered to be the main reason for the shift in Manila's South China Sea policy, as Ralf Emmers suggests, that "nationalist sentiments, especially if they are utilized by national governments as part of wider domestic political strategies, may constitute a formidable stumbling block toward the joint management of resources in the disputed waters".[96]

Although joint development in the South China Sea has made little real progress, Chinese scholars have never given up their studies of the possibilities. Given that there has existed a number of successful examples of joint development projects between Malaysia and Thailand, Malaysian and Vietnam in the Gulf of Thailand and joint development projects between China and Vietnam in the Beibu Gulf,[97] Chinese scholars have analysed the ASEAN countries' experiences in their joint development of maritime resources in the disputed waters, and proposed some possible joint development models, such as "joint development under joint management model", "joint development under proxy model", and "joint development under supernational management model".[98]

Hence, whether energy reserves in the South China Sea will become a source of cooperation or conflict largely depends on different considerations. From an economic perspective, joint development can help reach a better understanding of the real value of resources in the South China Sea and dilute sovereignty disputes while increasing mutual trust. Assuming that all sides are able to settle or shelve their competing territorial claims, the South China Sea is ripe for accelerated development and China is a natural market for the hydrocarbons that may be produced, then energy resources can become a key factor for stability and peace making, rather than a factor in conflict triggering. But this needs individual countries, especially big countries, to regard the energy potential in the South China Sea from a wider regional perspective, rather than being restricted to an individual viewpoint of sovereignty.

CONCLUSION

After twenty years of efforts, China–ASEAN friendly cooperation has made great achievements. Currently, apart from the South China Sea issue, there is no other fundamental obstacle in the bilateral relationship. Therefore, if the South China Sea disputes cannot be effectively managed or controlled, and eventually be resolved through a proper way, it is impossible to reach a higher level of good-neighbourly and friendly relationship between China and ASEAN, the relationship of "strategic partnership" between the two sides can neither be promoted to a higher level.

The current China–ASEAN relationship faces two "cruxes" in its development. On the ASEAN side, this is reflected in a "security dilemma" with ASEAN countries seeing China's rise as a potential threat to the region to a certain degree. On the China side, this is reflected in a "power anxiety" with China being anxious to gain external recognition of its peaceful rise. But to the contrast, this anxiety has instead aroused worries and concerns of its neighbours.[99] China has repeatedly stressed the peaceful nature of its development, but it seems that the effect of publicity is not satisfactory. As Singapore's late Minister Mentor Lee Kuan Yew once said in an interview:

> In my opinion, compared with the U.S., China may not be a benign hegemon after its rise. ... China frequently states that it would not be a hegemon. But since you are not ready to dominate, why are you so eager to declare to the world that you have no intention to seek hegemony? In contrast, we know that the U.S. is a hegemon, but it is a benign hegemon. I get along with them well. So why not let this benign hegemon continue to exist?[100]

This provides us a new thought to think of China's new policy to Southeast Asia and the South China Sea dispute.

We need to say that under Xi Jinping's administration, Beijing's strategy in the South China Sea issue is considered crucial to its ASEAN policy, driving it to seek more flexible and efficient measures. The withdrawal of the Haiyang Shiyou 981 oil rig demonstrated Beijing's quick learning capacity and boosted the people's confidence that Beijing's South China Sea policy will not be determined by any nationalist impulse. China conditionally welcomes a CoC arrangement based on a six-point guidance for China–ASEAN negotiations under Indonesian auspices.[101] If the status quo of the Spratlys in the 2000s is restored with the CoC, stand-offs would be prevented, serving China's core interest by providing precious breathing space for it to meet other priorities. This indicates that if everything remains status quo, South China Sea dispute is low in Beijing's diplomatic priority.

Chinese scholars have been actively exploring some policy suggestions to ease the tensions and resolve the disputes. For example, Zheng Yongnian believes that the urgent priority is to manage and control the South China Sea disputes. China should change its traditional way of thinking and search for a multilateral solution under a regional or even international framework.[102] Li Mingjiang suggests that China shift the basis of its Spratly claim. Rather than territorial or historic water based on the nine-dashed line map, China could claim instead the Spratlys as a line of islands and other land features covered by UNCLOS.[103] You Ji suggests that China could not put off those issues such as the nine-dotted line. It is necessary for China to move on to the following matters. Clarifying its South China Sea assertion and substantiating its insistence on the nine-dotted line,[104] putting forward its own roadmap for dispute resolution and promoting a new collective security concept. A new security concept can be flexible, including binding multilateral security mechanisms, multilateral forms, bilateral security consultations, and non-official academic security dialogues.[105] Moreover, taking into account the fact that the large-scale construction of islands will have an impact on the claims of ASEAN countries, it is appropriate for China to disclose information relating to the island's construction scale and its overall use.[106]

Some foreign scholars have suggested that China should adjust its terrestrial conception of bounded space. For example, Hans Dieter-Evers holds that the South China Sea is one of many "Mediterranean seas". It is typically bordered by a number of distinct states and surrounded by narrow outlets to oceans and other seas.[107] All Mediterranean seas experience periods of intensive trade relations, exchange of knowledge, economic prosperity, and the flowering of science, religion, and innovation. But unlike other Mediterranean seas and in spite of intensive trade relations and cultural exchanges, the South China Sea remains an entity only in name. Indonesians, Malaysians, and the Filipinos concentrated on their own seas, such as Sulu, Sulawesi, Java Sea, and the Strait of Malacca. Their concept is basically a maritime concept of free and undefined space; while the Chinese's view appears to be land based, and consider the South China Sea as a bounded territory and exclusive sovereign territory.[108] This adds difficulty to resolving of maritime disputes. One illustration is that "China has long called for joint development, but other claimant-states unease with Beijing's premise of 'indisputable sovereignty' has prevented any progress on the idea."[109] Dieter-Evers hopes that "the Chinese terrestrian conception of bounded space could be changed into a maritime conception of sharing resources to allow a solution of the claims to the South China Sea."[110] Indeed, as Chinese scholar Shao Jianping believes,

> as the largest country among the claimants and an emerging responsible power in Asia Pacific and the world, China should play a greater leadership role in promoting the joint development in the South China Sea, so that such joint development of maritime energy resources could lead to a longer-term prospect of territorial compromises and become an underpinning factor for the peace and stability in the South China Sea.[111]

Considering that decision-makers in China now often refer to scholars for advice, these suggestions might be internalized by Chinese leadership while it takes time for political conditions to mature.

Notes

1. Christopher Ward, "South China Sea on the Rocks: The Philippines' Arbitration Request", *East Asian Forum*, 21 April 2014.
2. Kane Jane, *The South China Sea Dispute: Prospect for Preventive Diplomacy* (Washington: United States Institute of Peace, March 1995).
3. U.S. Energy Information Administration, "The South China Sea", 7 February 2013, <http://www.eia.gov/countries/analysisbriefs/South_China_Sea/south_china_sea.pdf> (accessed 26 March 2015).
4. David Brown, "More Fuel to South China Sea Disputes", *Asia Times Online*, 12 March 2013.
5. R. Emmers, "Changing Power Distribution in the South China Sea: Implications for Conflict Management and Avoidance", Working Paper 183, S. Rajaratnam School of International Studies, September 2009.
6. On 2 May 2014, CNOOC towed its giant oil rig into disputed waters 240 km from Vietnam's coast. This was followed by each side accusing the other of attacking its vessels near the rig. In Beijing's perspective, it was a response to ExxonMobil's oil exploration in Block 143 on behalf of Vietnam since 2011.
7. Hans Dieter-Evers, "Understanding the South China Sea: An Explorative Cultural Analysis", *IJAPS* 10, no. 1 (January 2014).
8. Yan Xuetong, "Cong nanhai wenti shuodao zhongguo waijiao de tiaozheng" [From South China Sea Issue to the Adjustment of China Foreign Policy], *World Knowledge*, no. 1 (2012).
9. Richard Javad Heydarian, "China Casts Red Tape in South China Sea", *Asia Times Online*, 15 March 2014.
10. *China Customs Statistics 2012.*
11. Yan Xuetong, "Cong nanhai wenti shuodao zhongguo waijiao de tiaozheng" [From South China Sea Issue to the Adjustment of China Foreign Policy], *World Knowledge*, no. 1 (2012).
12. Pang Zhongying, "Dongmeng de waijiao xianjing" [Trape of ASEAN's Diplomacy], *Dongfang Zaobao*, 11 May 2012.
13. An ying-min, "Lun nanhai zhengyi quyu youqi ziyuan gongtong kaifa de moshi xuanze" [On the Mode of Jointly Developing Oil and Gas in the South China Sea], *Journal of Contemporary Asia*-Pacific, no. 6 (2011): 124–40.
14. *CNOOC Annual Report 2010*, <http://www.cnoocltd.com> (accessed 11 January 2014).
15. Stratfor Global Intelligence, "China Uses Deep-Sea Oil Exploration to Push Its Maritime Claims", <http://www.stratfor.com/analysis/china-uses-deep-sea-oil-exploration-push-its-maritime-claims> (accessed 14 May 2014).
16. Wang Kang-peng, "zhonghaiyou shiyou meng" [CNOOC's Oil Dream], <http://china5e.com/show.php?contentid-160243&page=3> (accessed 14 May 2014).

17. "Picking Apart Nationalist Rhetoric Around China's New Oil Rig", *The Wall Street Journal*, 11 May 2012.
18. Stratfor Global Intelligence, "China Uses Deep-Sea Oil Exploration to Push Its Maritime Claims", <http://www.stratfor.com/analysis/china-uses-deep-sea-oil-exploration-push-its-maritime-claims> (accessed 14 May 2014).
19. Wu Yin and Tang Jian, "Geo-economic Strategy in the South China Sea", in *China-Neighboring Asian Countries Relations: Review and Analysis*, edited by Li Xiangyan (Beijing: Social Science Academic Press, 2013), p. 153.
20. "Vietnam: PetroVietnam Finds More Oil at Bach Ho Field Offshore Vietnam", *Energy-Pedia News*, 29 June 2012, <http://www.energy-pedia.com/news/vietnam/vietsovpetro-finds-more-oil-at-bach-ho-field-offshore-vietnam> (accessed 14 May 2014).
21. Yu Xiangdong, "Yuenan quanmian haiyang zhanlue de xingcheng shulue" [A Brief Account of the Formulation of Vietnam's Comprehensive Ocean Strategy], *Journal of Contemporary Asia-Pacific Studies*, no. 5 (2008): 100–10.
22. Joseph Y.S. Chen, "Sino–Vietnamese Relations in the Early Twenty-first Century: Economics in Command?", *Asian Survey* 51, no. 2 (March/April 2011): 379–405.
23. Leszek Buszynski and Iskandar Sazlan, "Maritime Claims and Energy Cooperation in the South China Sea", *Contemporary Southeast Asia* 29, no. 1 (2007): 141–71.
24. *Annual Report of ONGC Videsh Limited 2010–11*.
25. Rup Narayan Das, "India in the South China Sea: Commercial Motives, Strategic Implications", *China Brief*, vol. xiii, issue 20 (10 October 2013).
26. Ibid.
27. Ananth Krishnan, "South China Sea Projects an Infringement on Sovereignty, Says China", *The Hindu*, 19 September 2011, <http://www.thehindu.com/news/international/article2468317.ece?css = print> (accessed 14 May 2013).
28. Ibid.
29. P.K. Ghosh, "Binding Vietnam and India: Joint energy exploration in South China Sea", Analysis, Observer Research Foundation, India, 17 December 2013.
30. IEA, *Southeast Asia Energy Outlook*, September 2013, p. 57.
31. *BP Statistical Review of World Energy June 2013*; IEA, *World Energy Outlook 2009*, p. 615.
32. "Philippines to Seek More Oil in West Philippine Sea", *Inquiry Global Nation*, 29 June 2012, <http://globalnation.inquirer.net/5034/philippines-to-seek-more-oil-in-west-philippine-sea> (accessed 14 May 2013).
33. Christopher Len, "Reed Bank: South China Sea Flashpoint", *Asia Times Online*, 3 June 2014.

34. Chi Fulin et al., "Development Program in the South China Sea and Construction of Hainan Strategic Base: Proposals for China's 'Eleventh-Five' Plan", *Review of Economic Research*, no. 51 (2005).
35. Wu Yin and Tang Jian, "Geo-economic Strategy in the South China Sea", in *China-Neighboring Asian Countries Relations: Review and Analysis*, edited by Li Xiangyan (Beijing: Social Science Academic Press, 2013), p. 154.
36. Sheldon W. Simon, "Conflict and Diplomacy in the South China Sea", *Asian Survey* 52, no. 6 (2012).
37. Cited in ibid.
38. Sam Bateman, "Managing the South China Sea: Sovereign is Not the Issue", *RSIS Commentaries*, no. 136 (29 September 2011).
39. Sheldon W. Simon, "Conflict and Diplomacy in the South China Sea", *Asian Survey* 52, no. 6 (2012).
40. Kyodo, "Abe Backs Up ASEAN on Maritime Security, Prods China", *Japan Times*, 30 May 2014.
41. Rober Ayson, "Japan Steals the Show at the Shangri-La Dialogue", *East Asia Forum*, 13 June 2014.
42. Ibid.
43. "Editorial: Government Must Boost Security, Economic Ties with ASEAN", *Daily Yomiuri*, 15 October 2012.
44. "Japan Muddies the Waters in the South China Sea", *China Daily*, 10 October 2011.
45. Aileen San Pablo-Baviera, "The Philippines in China's Soft Power Strategy", *ISEAS Perspective*, 3 June 2013.
46. David G. Wiencek and John C. Baker, "Security Risks of a South China Sea Conflict", in *Cooperative Monitoring in the South China Sea: Satellite Imagery, Confidence Building Measures, and the Spratly Island Disputes*, edited by John C. Baker and David G. Wiencek (Westport: Praeger Publishers, 2002), p. 54.
47. Cited in Ibid.
48. Global Indicators Database, <http://www.pewglobal.org/database/indicator/24/country/173/> (accessed 26 March 2015).
49. Renato Cruz de Castro, "Balancing Gambits in Twenty-First Century Philippine Foreign Policy Gains and Possible Demise?", in *Southeast Asian Affairs 2011*, edited by Daljit Singh (Singapore: Institute of Southeast Asian Studies, 2011), p. 240.
50. Jerry E. Esplanada, "Del Rosario Defines 3 Pillars of Foreign Policy", *Philippine Daily Inquirer*, 3 March 2011.
51. Dai Fang and Jin Shi-yong, "Anquan yu fazhan: feilibin dui hua zhengce yanjiu" [Security and Development: A Study on Philippines' China Policy], *Southeast Asian Affairs*, no. 3 (2009).
52. Hoang Oanh, "Vietnam's Deft Diplomatic Footwork on the South China Sea", East Asia Forum, 7 June 2014.

53. "U.S., Vietnam Start Military Relationship", *DefenseNews*, 1 August 2011, <http://www.defensenews.com/article/20110801/DEFSECT03/108010307/U-S-Vietnam-Start-Military-Relationship> (accessed 26 March 2015).
54. Lien Hoang, "US, Vietnam Inch Closer Together", *Asia Times Online*, 12 June 2012.
55. Hoang Oanh, "Vietnam's Deft Diplomatic Footwork on the South China Sea", *East Asia Forum*, 7 June 2014.
56. Rizal Sukma, "Indonesia–China Relations: The Politics of Re-engagement", *Asian Survey* 49, no. 4 (2009).
57. Leo Suryadinata, "South China Sea: Is Jakarta No Longer Neutral?", *Straits Times*, 24 April 2014.
58. On 12 March 2014, Indonesia's Commodore Fahru Zaini, assistant deputy to the chief security minister for defence strategic doctrine, was reported to have said that "China has claimed Natuna waters as their territorial waters. This arbitrary claim will have large impact on the security of Natuna waters." See Leo Suryadinata, "South China Sea: Is Jakarta No Longer Neutral?", *Straits Times*, 24 April 2014.
59. Ann Marie Murphy, "Jakarta Rejects China's Nine-Dash Line", *Asia Times Online*, 3 April 2014.
60. Ibid.
61. "Indonesia Seeks for Road for South China Sea", *South China Morning Post*, 26 September 2012.
62. Wangsen and Yang Guanghai, "Dongmeng daguo waijiao zai nanhai wenti shang de yunyong" [Implementation of ASEAN's Big Power Diplomacy in the South China Sea Issue], *Journal of Contemporary Asia-Pacific Studies*, no. 1 (2014): 35–57.
63. "Chongdie wenti: zhongguo bei huyu ying gengjia jinshen" [Overlapping Issue: China Was Called for Being More Cautious], *Herald* (Malaysia), 16 June 2011.
64. Liselotte Odgaad, "The South China Sea: ASEAN's Security Concerns about China", *Security Dialogue* 34, no. 1 (March 2003).
65. Chinese Premier Wen Jiabao's speech at the China–Malaysia Trade and Investment Cooperation Forum, 29 April 2011, <http://news.sina.com.cn/c/2011-04-29/020622377866.shtml> (accessed 14 May 2014).
66. Carlyle A. Thayer, "Can ASEAN Respond to the Chinese Challenge", *YaleGlobal Online*, 18 March 2014, <http://yaleglobal.yale.edu/content/can-asean-respond-chinese-challenge> (accessed 14 May 2014).
67. Ibid.
68. "Malaixiya zhuanzhu quanqiu anquan" [Malaysia Focuses on Global Security], *Herald* (Malaysia), 4 June 2011.
69. Wangsen and Yang Guanghai, "Dongmeng daguo waijiao zai nanhai wenti shang de yunyong" [Implementation of ASEAN's Big Power Diplomacy in

South China Sea Issue], *Journal of Contemporary Asia-Pacific Studies*, no. 1 (2014): 35–57.

70. For instance, instead of engaging ASEAN and other important Asian markets, either individually or through attempting to join ASEAN-centred RCEP – the preferred economic engagement vehicle for most of the East Asian countries – the U.S. has focused on the TPP, which involves only four of ten ASEAN countries and for which only seven of ten are eligible.
71. *China's Customs Statistics Yearbook 2011*.
72. Aileen San Pablo-Baviera, "The Philippines in China's Soft Power Strategy", *ISEAS Perspective*, 3 June 2013.
73. Renato Cruz de Castro, "Balancing Gambits in Twenty-First Century Philippine Foreign Policy: Gains and Possible Demise?", in *Southeast Asian Affairs 2011*, edited by Daljit Singh (Singapore: Institute of Southeast Asian Studies, 2011), p. 240.
74. Jose L. Tongzon, "Trade Policy in the Philippines: Treading a Cautious Path", *ASEAN Economic Bulletin* 22, no. 1 (2005).
75. Ministry of Commerce of China, *2012 Statistical Bulletin of China's Outward FDI*.
76. Quoted from Hamza Mannan, "Kerry Missing the Boat on Asia", *Asia Times Online*, 14 January 2014.
77. IMF, *Direction of Trade Statistics Yearbook 2012*.
78. T. Davenport, "Joint Development in Asia: Some Valuable Lessons Learned", in *Maritime Energy Resources in Asia: Legal Regimes and Cooperation*, edited by Clive Schofield, NBR Special Report #37 (February 2012): 140.
79. Ralf Emmers, *Resource Management and Contested Territories in East Asia* (Basingstoke: Palgrave Macmillan, 2013), p. 60.
80. Ibid., p. 67.
81. A. San Pablo-Baviera, "The South China Sea Disputes: Is the Aquino Way the 'ASEAN Way'?", *RSIS Commentaries*, 5 January 2012.
82. Ralf Emmers, *Resource Management and Contested Territories in East Asia* (Basingstoke: Palgrave Macmillan, 2013), p. 66.
83. Christopher Len, "Reed Bank: Next Flashpoint for China and the Philippines in the South China Sea?", *Policy Brief* (Institute for Security & Development Policy), no. 153 (26 May 2014).
84. Philip Andrews-Speed and Roland Dannreuther, *China, Oil and Global Politics* (Abingdon: Routledge, 2011), p. 147.
85. Ralf Emmers, *Resource Management and Contested Territories in East Asia* (Basingstoke: Palgrave Macmillan, 2013), p. 61.
86. Barry Wain, "Manila's Bugle in the South China Sea", *Far East Economic Review*, January/February 2008.
87. You Ji, "China's Civil-Military Strategies for South China Sea Dispute Control", *EAI Background Brief*, no. 1002 (25 February 2015).

88. Ibid., p. 66.
89. Cha Wen, "Feilibing nanhai zhengce zhuanbian beihou de guonei zhengzhi yiinsu" [Domestic Political Factors Behind the Philippines' Shift in Its South China Sea Policy], *Journal of Contemporary Asia-Pacific Studies*, no. 5 (2014): 120–39.
90. Ibid.
91. Ibid.
92. Evan S. Medeiros et al., *Pacific Currents: The Responses of US Allies and Security Partners in East Asia to China's Rise* (Santa Monica: Rand Corporation, 2008).
93. International Crisis Group, "Stirring Up the South China Sea (II): Regional Responses", 2012, p. 21.
94. Evan S. Medeiros et al., *Pacific Currents: The Responses of US Allies and Security Partners in East Asia to China's Rise* (Santa Monica: Rand Corporation, 2008).
95. Cha Wen, "Feilibing nanhai zhengce zhuanbian beihou de guonei zhengzhi yiinsu" [Domestic Political Factors Behind the Philippines' Shift in Its South China Sea Policy], *Journal of Contemporary Asia-Pacific Studies*, no. 5 (2014): 120–39.
96. Ralf Emmers, *Resource Management and Contested Territories in East Asia* (Basingstoke: Palgrave Macmillan, 2013), p. 61.
97. Shao Jianping, "Ruhe tuidong 21 shiji nanhai gongtong kaifa – dongnanya guojia jingyan de sijiao" [Promoting Joint Development in the South China Sea: Lessons from Southeast Asian Countries], *Journal of Contemporary Asia-Pacific Studies*, no. 6 (2011): 142–58.
98. An Yingmin, "Lun nanhai zhengyi qu youqi kaifa ziyuan gongtong kaifa de moshi xuanze" [A Discussion of Possible Models for Joint Development of Gas and Oil Resources in Contested Waters in the South China Sea], *Journal of Contemporary Asia-Pacific Studies*, no. 6 (2011): 124–40.
99. Cao Yun Hua and Ju Hailong, eds., "Nanhai diqu xingshi baogao 2012–2013" [South China Sea Regional Report: 2012–2013].
100. Hang Fu Guang, "Li Guangyao: xin jia po laiyi shengcun de ying daoli" [Lee Kuan Yew: Singapore's Last Reason on Which to Survive] (Singapore: Strait Times Press, 2011), p. 287.
101. In his Southeast Asian trip in May 2013, Foreign Minister Wang Yi reiterated Beijing's willingness to work with ASEAN for a mutually acceptable CoC. But Beijing insists that the CoC is for crisis management; it should not touch on EEZ demarcation.
102. Adopted from Zhen Yong, "Guonei nanhai wenti zhongsu" [Review of the South China Sea studies in China], *Journal of Contemporary International Relations*, no. 8 (2012).
103. See Alice D. BA, "Staking Claims and Making Waves in the South China Sea: How Troubled are the Waters?", *Contemporary Southeast Asia* 33, no. 3 (2011).

104. China's Civil-Military Strategies for South China Sea Dispute Control, *EAI Background Brief*, no. 1002, 25 February 2015.
105. Zhang Biwu, "Chinese Perceptions of U.S. Return to Southeast Asia and the Prospect of China's Peaceful Rise", *Journal of Contemporary China* 24, no. 91 (2015): 176–95.
106. Xue Li, "China and the U.S. did not enter the South China Sea military conflict countdown", FT Chinese network, 27 May 2015.
107. Hans Dieter-Evers, "Understanding the South China Sea: An Explorative Cultural Analysis", *IJAPS* 10, no. 1 (January 2014).
108. Ibid.
109. Jeffrey Ordaniel, "The geopolitical stakes of the 2016 Philippine election", *The Diplomat*, 28 May 2015.
110. Ibid.
111. Shao Jianping, "Ruhe tuidong nanhai gongtong kaifa? dongnanya guojia jingyan de sikao" [How to Promote Joint Development of the South China Sea? Perspectives from the Experiences of Southeast Asian Countries], *Journal of Contemporary Asia-Pacific Studies*, no. 6 (2011).

7

CONCLUSION

The twenty-first century is known to be the Asia century. The coming decades will see global economic growth increasingly being generated by Asian emerging economies. ASEAN is seeking to build political security and sociocultural communities as well as the economic community. With the rising population and economic growth resulting in an increasing share of global gross domestic product (GDP), Southeast Asia is seeing rapid growth in its energy demand and is consequently shifting the centre of gravity of world energy demand to Asia, along with its neighbours of China and India. These countries will account for an increasing share of international trade, investment flows, and energy resource demand. Such dramatic and rapid changes and shifts have created challenges in providing for energy resource supply, energy security, and environmental managements, as well as challenges to the international relations in the region.

COOPERATION OR CONFLICTS?

The previous chapters have analysed that, from an economic point of view, Asia is increasingly regarded as the most important region in the world. Indeed, it is widely thought to be the next most important focal point of the world economy that may soon replace the United States as the engine of global economic growth. The rise of China, the ASEAN Community, and other Asian countries like India economically has prompted East Asia to become the world's new economic centre of gravity and created a great impact on the world energy resource market, and will hence accelerate the changes in the world energy system, accelerating the energy shift from west to east.

In the background of rapid economic growth and increasing energy resource shortage, China's broader energy security strategy has been to pursue supply diversity, to find more and develop more offshore oil and gas resources in Southeast Asia. China has the intention to invest more on natural gas sectors in some ASEAN countries, such as Indonesia, Myanmar, and Malaysia. China expanded its foreign direct investment (FDI) to Southeast Asia after the global financial crisis in 2008, and China–ASEAN energy cooperation has developed to a new level, extending from energy trade to energy resource exploration and related infrastructure-building.

While some momentum exists for continued cooperation and energy cooperation has largely broadened China–ASEAN bilateral relations, some factors are pushing the region toward competition and conflict. The rise of China's outward FDI, its resilience during the global financial crisis and its continuing rapid growth have raised questions about the "China model" of investment abroad. Some analysts have argued that China's "going out" strategy is mercantilist in nature, and in Southeast Asian countries, this strategy has contributed to mounting distrust and concerns in local communities (such as in Myanmar and Indonesia, Chapters 4 and 5). Hence, it is believed that China's request for energy resources may become the spark for regional conflicts and international instability.

Nevertheless, to view this energy resource quest strategy as a purely Chinese phenomenon would be unfair. There is evidence showing the spread of resource nationalism across Asia, both at national and international levels, which fuels long-standing rivalries and possible conflicts. At a national level, there is rising resource nationalism in host countries to control natural resources and maximize rent appropriation to benefit national development. A resource-nationalist orientation might not cause disputes between governments and foreign investors, but it certainly makes existing tensions and concerns more difficult to resolve (Chapter 4).

At the international level, in fact, resource security strategies are not without historical precedent in Northeast Asia. During the 1960s and 1970s, the Japanese government offered financial and diplomatic assistance to its industrial corporations to sponsor the development of new mining firms in Latin America and Southeast Asia. These mining firms were connected to the Japanese economy through long-term contracts and investment ties, which guaranteed Japan's resource security by ensuring their output was preferentially supplied to the Japanese market.[1] South Korean heavy industrialization in the 1970s also saw its government launch a similar resource diplomacy programme to strengthen intergovernmental ties with its foreign energy suppliers.[2] Similar to China's programme, Japan and Korean strategies have also called for national firms to make security-oriented investments in overseas resource projects, providing government assistance to national firms to acquire ownership, and ultimately control, of overseas resource projects.

More recently, South Korea developed new resource security policies and announced both a National Energy Plan and an Overseas Resource Development Plan in 2004. The Japanese government followed soon thereafter, issuing a New National Energy Strategy in 2006 and a set of Guidelines for Securing Natural Resources in 2008.[3] Similar to China's programme, Japanese and South Korean strategies have also called for national firms to make security-oriented investments in overseas resource projects, providing government assistance to national firms to acquire ownership, and ultimately control, of overseas resource projects.

In the case of China, the Chinese government extends a variety of incentives and means of support to its companies, and China's rapid growing foreign exchange reserves allow state-owned financial institutions to underwrite overseas investment expansion. But compared with Japan and other developed countries, China launched its ASEAN resource diplomacy much later and the foreign aid disbursement, with its potential role in securing resource contracts, is still quite limited in size.[4] Yet little attention has been paid to the strategies of other resource consumers in the region, or whether China's approach might share common causes and implications with those of Japan and South Korea. Actually, it is the Chinese companies' misbehaviours such as lacking corporate social responsibilities that have caused international and host countries' concerns.

Then what might have caused those misbehaviours by Chinese outward foreign direct investments (OFDI) companies? We need to say that most of these Chinese firms were newcomers, and did not know enough about local laws, regulations, and cultures. They face a steep learning curve in their OFDI. A survey in 2010 shows that more than 80 per cent of Chinese OFDI firms began investing abroad after 2000, while only 4 per cent of them had outward investments before 1990.[5] More importantly, most of these firms have loosely abided by the law before they went out, and the Chinese legal system and requirements on business ethics, especially those with regard to environmental protection are immature compared to that of Japan, South Korea, and other developed economies. It was thus possible for them to keep doing what they were used to in foreign markets.

So in addition to better understanding local conditions, Chinese firms should first strengthen corporate transparency and compliance so as to win trust and positive perceptions from local communities. Evidence shows that China has been improving in this regard. For example, the State-owned Assets Supervision and Administration Commission of the State Council of China issued the "Guideline on SOEs [state-owned enterprises] Fulfillment of Corporate Social Responsibility (CSR)" in late 2007 and asked SOEs, including big extractive companies like China National Petroleum Corporation (CNPC), China Petroleum and Chemical Corporation (Sinopec), and China National Offshore Oil Corporation (CNOOC), to issue CSR reports to disclose information relating to their social responsibilities to stakeholders and the public in general, and receive their supervision. One could go further and see if Chinese firms investing in Southeast Asia and other regions will sign up the Extractive Industries Transparency Initiative (EITI) which complements China's SOE's reporting requirements and is aimed at improving transparency on the activities of National Oil Companies (NOC) overseas.[6]

While it is accepted that energy resources can play a role in conflict triggering, the fact that energy resource rivalry might also create cooperation opportunities should not be ignored. As China has followed its growth path, its concern about resource security has intensified and the Chinese government has encouraged FDI as a means of securing resources. Chinese capital will flow more intensively into the developing world, especially its neighbouring countries. This has, in fact, opened large-scale development opportunities in the resource sectors, especially natural gas sectors in Southeast Asia. Moreover, China is also motivated by the prospect of becoming a leader in emerging energy technologies. China has become the world's largest producer of solar panels and wind turbines. And the race is on for global leadership in transformative energy technologies such as electric cars and smart electricity grids.

On the side of ASEAN, Southeast Asia is richer in natural gas than in oil. At current levels of production, ASEAN's proven reserves of gas would sustain production for another thirty-five years. However, the gas development in Southeast Asia also meets some impediments, as major gas producers such as Indonesia and Malaysia face difficulties in allocating supply between domestic demand and exports. ASEAN needs significant investment to related infrastructures so as to bring forward the projected amount of oil, gas, and coal production. Financing investment in energy development and energy-supply infrastructures has always been a big challenge for ASEAN countries.

Given the importance of China's capital, technology, and big market, China will remain as ASEAN's major energy cooperation partner. More importantly, China has maintained close relations with ASEAN not only in the energy field but also in international politics, economic cooperation, and business transactions. As ASEAN's importance for international markets increases with various energy challenges emerging in the region, China's deepening of relations with ASEAN will be of growing strategic importance.

However, it is important to note that given the disquiet and uneasiness among Southeast Asians about the nature and impact of the bilateral relationship are still prevailing, the extent to which China's energy resource cooperation with ASEAN countries such as Indonesia and Myanmar will develop depends on the efforts of both sides. On the one hand, it depends on whether domestic constituencies in these countries, primarily labour unions, organized businesses, and civil society formations, feel that their concerns about China's influence on their political economy are being addressed. Particular areas of concern are the continuing impact of Chinese FDI on local jobs and environment, the erosion of the competitive advantage of domestic companies (in local and regional markets) by the growing presence of Chinese companies, the skewed trade relations, and perceptions that expanding commercial relations have exerted a baleful influence on these countries' foreign policy.

On the other hand, it also depends on whether the host countries can depoliticize some high-profile Chinese deals and rationalize their public debates about the risks and benefits of Chinese investment. Nevertheless, on the China side, Beijing's willingness to comply with Indonesia's quest to correct the unbalanced trade profile between the two countries by reducing textile exports and urging Chinese firms to expand investment in the country's manufacturing sector; and Beijing's promise to make corporate social responsibility programmes an integral component of Chinese SOEs' operation in Myanmar suggest that Beijing is well attuned to the sensitivities of Southeast Asian countries' internal politics.

WORKING TOGETHER FOR A NEW WORLD ENERGY ORDER

At the international level, China and ASEAN countries can work together with the United States, Japan, and other Western countries to shape a new world energy order. The existing international trade and energy security institutions are the product of a post-World War II order that reflected a decidedly different time and place. The emerging China and ASEAN countries as new global players with different cultures, business practices, and foreign policy agendas have highlighted growing questions about the relevance and effectiveness of the existing energy system, and has required Western countries and these institutions to adapt to changing circumstances and create a new international energy order. The International Energy Agency (IEA), for example, is grappling with how to draw in major new consumers like China, India, and ASEAN countries that are not members of the Organisation for Economic Co-operation and Development (OECD). The United States and other Western countries need to accept and accommodate China and some ASEAN countries like Indonesia as emerging energy giants to mitigate possible conflicts, and then they can work together and address energy-related political and strategic issues.

The rising dependence of China and ASEAN countries on international oil and gas supplies requires a new set of relations with the United States and with each other. Given the facts of China's and ASEAN's inevitable rise as significant players and regional rivalries, the United States as a dominant power needs to adapt and adjust to the new international conditions and dynamic developments in the global oil markets, and play a role guiding the growing involvement of China and ASEAN countries within a broad cooperative framework. If the United States can adjust to the new international conditions and dynamic changes in the global and regional markets, making some uncomfortable concessions instead of exaggerating the differences, dangers and controversies, major conflicts with China for energy security issues might well be avoided.

Washington has the opportunity to shape the international environment in a way that can assist China and ASEAN countries to achieve energy efficiency and adjust to the world energy markets. They share a basic strategic interest in regional stability and the free flow of energy resources. These common interests create a platform for cooperation that can enhance not only security in the Middle East, Africa and other oil rich regions, but also Sino–U.S.–ASEAN relations more generally. This partnership can be promoted on the following two dimensions.

Firstly, the United States as a superpower can play its role to lead China and ASEAN countries onto the right track by, for instance, taking

China and ASEAN emerging countries to the IEA, so as to turn China's unilateral energy policy into a multilateral one. At the current stage, China is "partner country" in IEA. One criterion for full membership in the IEA is that a country has to reach the level of per capita GDP for membership in the OECD. "If the IEA redesigned its admission criteria and brought China into the fold, it may help dampen Chinese nationalists' complaint about status in the world, and challenge China to shoulder required responsibilities."[7] Indeed, it is essential for China and ASEAN countries to be further integrated into the global oil and gas market system by being provided with more opportunities to participate in the decisions on the rules governing that market and share information on world oil and gas markets. China and ASEAN countries should be brought into coordination with the existing IEA energy security system to assure that their interests and energy security will be protected in the event of turbulence, and to ensure that the system works more effectively so that they can follow and abide by the international rules more closely.

Secondly, China remains the second largest emitter of carbon dioxide after the United States. In recent historical terms, no major economies have managed to decouple economic growth from heavy emissions at early stages of their development. China and ASEAN emerging countries are no exceptions. Developed countries can pursue technical collaboration with China and ASEAN countries for energy conservation and efficiency, for using clean and renewable energy sources, for seeking safe and alternative energy, and for reducing oil dependency by jointly developing the next generation of energy sources and a more stable global and regional energy order.

Large-scale and long-term cooperation between the United States and China and ASEAN countries in the energy and environmental sectors has profound strategic significance. It will benefit all parties economically; it will guide China's and ASEAN countries' energy policies in a more environmental friendly directions; and, more importantly, it will modify Beijing's foreign policy along a more peaceful and less confrontational path, thus serving the comprehensive security interests of the United States, Japan, European Union, and the rest of the world.

To sum up, it is in the interests of the United States, Japan, and other Western countries to establish a constructive relationship with Beijing and ASEAN countries in energy resource security area. This will help China and ASEAN countries craft more conservationist development strategies and adapt to the world energy system.

WORKING TOGETHER FOR A NEW REGIONAL ORDER

We need to say that energy resource cooperation between China and ASEAN countries is only part of China–ASEAN relations. It can promote and influence bilateral relations, as well as be affected by them. China's energy expansion can have mixed impacts on Southeast Asia. Actually, the more important aspects of such impacts arise not from China's surging demand for energy, but more from what policies and approaches it adopts in the region and beyond to address its energy security.

It is true that there exist some friction and suspicions between China and ASEAN countries, and this is hardly surprising in view of the United States' global position and its rebalancing strategy or "return to Asia" policy, China's rapid rise and the rising role of the ASEAN community in the region. This is particularly reflected in the South China Sea disputes. Indeed, China's maritime challenge is multi-faceted, involving strategic rivalry with the U.S. and maritime disputes with its neighbours. To a large extent the territorial and maritime disputes in the South China Sea are getting more complicated and multilateralized when Washington gets involved. But for the perceived threat from the United States as an obstacle to China–ASEAN relations, as Chinese scholar Zhang Biwu says, China should take a relaxed view. The United States is able to increase its influence there only because local people welcome its presence. In the 1990s when Filipinos resented the U.S. presence, the U.S. had to close its naval and air bases there. With increased U.S. presence in Southeast Asia, people in this region might also be more comfortable with an increase in Chinese influence.[8] Therefore, China should probably regard the increased U.S. influence as an opportunity rather than as a threat to China's interest in Southeast Asia.

Recent years have seen the international relations of Asia in a period of rapid and profound transition in the context of dynamic geopolitical and economical change. The United States may still be the predominant power that provides the security public goods that have facilitated the rapid economic growth of most of the Asian states, and continue to play an essential role of setting norms and rules for global commerce, but U.S. predominance is being challenged — in part because of the rise of China and other emerging powers and in part because of the relative decline of the United States itself and its recent rebalancing policy in Asia.

Although both India and Japan are seeking to enhance their influences in Southeast Asia, it is the rise of China that is contributing most to the dynamics of changes in Southeast Asia and beyond. China's process of economic reform began much earlier than that of India, and its impact on the Southeast Asian and global economy far exceeds that of India. Combined with deft economic diplomacy, China's political weight has increased dramatically and the influence is bound to further increase. China has also been rapidly modernizing its military forces, and this reality is beginning to stir concerns in its Southeast Asian neighbours and is seen as a potential challenge to the United States' dominant position in East Asia and beyond.

ASEAN countries have played a significant part in mediating relations between the major powers (China and Japan, China and the United States) and shaping a new regional order in Asia. As the ASEAN Community is in the process of being established, ASEAN countries are becoming more coherent and united. ASEAN has determined the distinctive mode of the operations of the regional and economic institutions. ASEAN's norms of procedure by consensus and non-interference have not only facilitated China's participation in regional multilateral institutions but have formed the basis for the operations of all the other regional institutions. Moreover, in the absence of trust between major powers such as China and Japan in the region, ASEAN can and has filled the void as a leader to address regional problems and lead regional economic cooperation.

Now the dynamic relations between China and the United States are raising new questions and challenges among Washington's allies and friends in Southeast Asia especially the South China Sea, resulting in some divides in ASEAN. China has adjusted its South China Sea policy and used its enhanced naval power in pursuit of its long-standing territorial claims in the East and South China seas, which has led to some external players increasingly getting involved in the South China Sea disputes. As the United States has been getting more involved in the South China Sea disputes, some ASEAN countries like Malaysia and Singapore prefer to see the United States adopt a "more modest role" that would not require them to choose sides with the United States or China. They prefer and seek to maintain good relations with both the United States and China.

Nevertheless, Asian countries' reliance on the United States for energy shipment security and regional stability will create greater incentives for them to be cooperative in U.S. diplomacy, or at least not to disrupt it. China is not the exception, since around 70 per cent of its imported oil originates in Africa and the Middle East, the only way to inexpensively transport it is by sea. Inadequate naval forces guarantee China will continue to depend on the U.S. Navy to protect its oil and gas trade. The territorial disputes surrounding the South China Sea islands do not change that. Chinese leaders have said that while they are determined to protect their national interests, they are capable of holding back when the aggressive pursuit of those interests may bring them to a decisive confrontation or conflict with the United States. There are clear signs showing that the current Chinese leaders are aware of the importance of avoiding the South China Sea disputes as it would affect China–ASEAN relations. China and ASEAN countries have common interests with the United States in keeping the South China Sea lanes safe and regional stability high. These common interests may create a platform for cooperation that can enhance not only political stability in those regions, but also China–ASEAN and other big powers relations more generally.

The relationship between China and the United States is one between a rising power and an established dominant power. Competition in Southeast Asia is inevitable although "the balance of interest in the region strongly favours China because the various diplomatic and territorial quarrels roiling East Asia are of much greater salience and concern to China than to the U.S."[9] While it has become difficult for the U.S. to hold its primacy in the region, China cannot be a single power of domination in the region either. The two powers will have to develop a clearer mutual understanding and greater mutual acceptance, and work together to maintain a balance of power in the region to limit strategic rivalry.[10]

ASEAN is at the centre of this and has been able to establish platforms that could play a supplementary role in channeling U.S.–China relations in more predictable and constructive directions, because U.S. and Chinese interests intersect in Southeast Asia; ASEAN is a relatively neutral body friendly to both; and both the United States and China find it useful to use these ASEAN-created platforms.[11] In other words, Southeast Asia can be a theatre where China and the United States can introduce a new framework for cooperation that recognizes the reality of the two countries' strategic considerations. This minimal role is useful although will not be the decisive factor.

The interaction process would be prolonged and challenging. It requires that both China and the United States follow a balanced and pragmatic policy dealing with each other in the region. For the United States, the challenge is how to reassure China of its alliance policy in Asia while being more open-minded to building a regional architecture that can accommodate the interests of China and the rest of Asia. As Stephen Hadley suggests, whether Washington is willing to counsel restraint to its friends and allies in the Asia-Pacific region or even urge them to make efforts to compromise with China on territorial dispute issues will be a touchstone for building the new model of U.S.-China relations.[12] China, on the other hand, needs to find more innovative ways to reassure Washington and its Asian neighbours of its peaceful rise and play a more responsible role contributing to the regional economy and especially politics and security. This interaction can thereby narrow the trust gap between the two countries, and eventually help build a more balanced regional order and a new type of China–U.S. relationship through cooperation in Southeast Asia and beyond.

Notes

1. Ozawa Terutomo, *Multinationalism, Japanese Style: The Political Economy of Outward Dependency* (Princeton: Princeton University Press, 1979).
2. Lee Min Yong, "Securing Foreign Resource Supply: Resource Diplomacy of South Korea", *Pacific Focus* 3, no. 2 (1988): 79–102.
3. Ministry of Economy Trade and Industry of Japan, *Guidelines for Securing Natural Resources*, March 2008, <http://www.meti.go.jp/english/newtopics/data/pdf/080328Guidelines.pdf> (accessed 1 April 2015).
4. For example, from 2010 to 2012, China appropriated a total of US$14.4 billion for foreign assistance in three types: grants (aid gratis), interest-free loans, and concessional loans. See *China's 2014 Foreign Aid White Paper*, <http://news.xinhuanet.com/english/china/2014-07/10/c_133474011.htm> (accessed 1 April 2015). By contrast, in 2012 the Japanese government allocated a total official development assistance (ODA) of US$18.7 billion. See Japan's ODA White Paper 2013, <http://www.mofa.go.jp/policy/oda/white/2013/pdfs/all.pdf> (accessed 1 April 2015).
5. Quoted from Bijun Wang and Rui Mao, "Overseas Impacts of China's Outward Direct Investment", *Asian Economic Policy Review* 9, issue 2 (2014): 227–49.
6. The Extractive Industries Transparency Initiative (EITI) was founded at a conference in London in 2003. It is an international organization and maintains a standard, assessing the levels of transparency around countries' oil, gas, and mineral resources. This standard is developed and overseen by a multi-stakeholder board, consisting of representatives from governments, extractives companies, civil society organizations, institutional investors, and international organizations. EITI Standard is implemented in forty-eight countries. It consists of a set of requirements that governments and companies have to adhere to in order to become recognized as "EITI Compliant". The Chinese government has expressed interest and support to EITI.

7. Zha Daojiong, "Chinese Energy Security Vulnerabilities", paper presented at RSIS workshop on China and Non-Traditional Security: Global Quest for Resources and Its International Implications, 31 October 2014.
8. Zhang Biwu, "Chinese Perceptions of US Return to Southeast Asia", *Journal of Contemporary China* 24, no. 91 (2015): 176–95.
9. Suisheng Zhao, "A new model of big power relations? China–U.S. Strategic rivalry and balance of power in the Asia-Pacific", *Journal of Contemporary China* 24, no. 93 (2015): 377–97.
10. Ibid.
11. Bilahari Kausikan, "The Idea of Asia", *Straits Times*, 8 November 2014.
12. Stephen J. Hadley, "Asia-Pacific Major Power Relations and Regional Security", World Peace Forum, Beijing, 21 June 2014, <http://carnegieendowment.org/2014/06/21/asia-pacific-major-power-relations-and-regional-security> (accessed 29 September 2014).

INDEX

S

ABOUT THE AUTHOR

ZHAO Hong is Senior Fellow at ISEAS–Yusof Ishak Institute, Singapore. Prior to joining the Institute, he was a senior research fellow at the East Asian Institute (EAI), National University of Singapore; a professor at the Research School of Southeast Asian Studies, Xiamen University, where he taught International Political Economy, Big Power Relations, and Southeast Asian Economy. His latest published book and papers include: *China and India: The Quest for Energy Resources in the 21st Century* (2012), "China's Dilemma on Iran: Between Energy Security and a Big Responsible Country" (2014). His research mainly looks at the political economies of China, Southeast Asia, and South Asia; economic growth and energy security in Asia; Asia-Pacific regionalism and economic community.

CPSIA information can be obtained at www.ICGtesting.com
Printed in the USA
BVOW06s1946131016

464985BV00010B/71/P